THE GREEN-EYED DRAGONS

and Other Mathematical Monsters

David Morin

Harvard University

© David Morin 2018
All rights reserved

ISBN: 978-1719958370

Printed by KDP Print

Additional resources located at:
www.people.fas.harvard.edu/~djmorin/book.html

Cover image: Illustration by Maki Naro
makinaro.com

Preface

This book is a collection of 57 *very* difficult math problems I have compiled over the years. The collection started long ago in graduate school. Every now and then during those years, one of my fellow graduate students would come into the office and say, "Hey, I just came across a new problem, have you heard of this one?..." Whenever research was going slowly, it was always comforting to have an interesting problem to puzzle over! Many of those problems eventually found their way to an old "Problem of the Week" webpage of mine (www.physics.harvard.edu/academics/undergrad/problems). After letting the problems sit there for a while, I finally (re)polished up a number of them, added some new ones, and created this book.

The book is written for anyone who (a) enjoys pondering difficult problems for great lengths of time, and (b) can tolerate the frustration of not being able to figure something out. This book isn't for the faint of heart. If you use it properly (that is, without looking at the solutions too soon; see the comments below), you *will* get frustrated at times, and you *will* pull out a few hairs. But just because "No pain, no gain" is a cliche, that doesn't mean it's not true!

Chapter 1 contains the problems (57 in all), Chapter 2 gives some hints, and Chapter 3 presents the solutions. There is also an appendix on Taylor series. The hints in Chapter 2 are fairly minimal, so don't expect a problem to be easy after looking at a hint. For better or for worse, I decided not to rate the problems with a difficulty level.

For most of the problems, algebra is the only formal prerequisite. But a few require calculus, if you want to steer clear of those. They are: Problems 8, 11, 12, 37, 39, 40, 42, 44, 45, 52, and 53. This list doesn't include ones that make use of the results given in Problems 52 and 53 (even though calculus is an ingredient in those). It also doesn't include problems that use (but don't require the derivation of) a Taylor series from the list given in the appendix, because the *use* of a Taylor series involves only algebra; see the appendix for more on this.

The problems are mostly classics that have been around for ages. The solutions (for the most part) are mine, although I'm sure that every one of them has appeared countless times elsewhere. The problems are divided into four categories: General, Geometry, Probability, and Foundational. The probability section is the longest of the four. If you want to review some concepts from probability and combinatorics (binomial coefficients, expectation value, etc.), you may want to take a look at my book *Probability for the Enthusiastic Beginner*. A few problems from that book also appear in this one.

The Foundational problems contain results that are useful in other problems throughout this book (mostly in the probability section). The results from Problems 52 (Stirling's Formula) and 53 (A Handy Formula) are the most useful, so you might want to do those (or at least note their results) before diving into the probability problems. If a given problem requires a result from a foundational one, I will usually make a reference to that in the statement of the problem. I chose to put the foundational problems last in the book instead of first, because they tend to be of the more technical math type, and I didn't want readers to assume that those problems had to be done first. You can think of them sort of as appendices. Overall, there is no preferred order for doing the problems. They are arranged somewhat randomly within each section, so you can jump around and tackle whichever problem looks appealing on a given day.

The solutions often contain extensions/variations of the given problems. You can treat these as additional problems that are waiting to be solved. Just set the book aside and try to derive/prove the result yourself, without looking at how I did it. That way, there are even more than 57 problems in the book!

The most important advice I have for using this book is:

> **Don't look at the solutions too soon!**

The problems are designed to be brooded over for a while. If you look at a solution too soon and thereby remove any chance of solving things yourself, then the problem is gone forever. It's never coming back. There are only so many of these classics in the world, so don't waste them by looking at the solution without thinking about the problem for a *long* time. How long? Well, if you can't solve a problem, wait at least a week before looking at the hint. If that doesn't do the trick, then wait at least a month before looking at the solution. Actually, even a month is probably too short. There's really no hurry. Just move on to another problem; there are lots of them. As long as there are other problems to work on, there's no need to look at any solutions. You can be pondering many at a time.

If you do eventually need to look at a solution (after at least a month), you should read only one line at a time, covering up the page with a piece a paper, so that you don't accidentally see too much. As soon as you read enough to get a hint, set the book aside and try to work things out. That way, you'll still be able to (mostly) solve the problem on your own. Repeat as necessary, with a week between peeks at the solution. You will learn a great deal this way. If you instead head right to the solution and read it straight through, you will learn very little.

A few informational odds and ends: This book contains many supplementary remarks that are separated off from the main text; these end with a shamrock, ♣. The figures were drawn with Adobe Illustrator. The numerical plots were generated with Wolfram Mathematica. I often use an " 's" to indicate the plural of one-letter items (like 6's on dice rolls). I refer to the normal distribution by its other name, the "Gaussian" distribution. I am occasionally sloppy with the distinction between "average value" (dealing with past events) and "expected/expectation value" (dealing with future events). And in quotients such as $a/(bc)$, I often drop the parentheses and just write a/bc; I do *not* mean $(a/b) \cdot c$ by this.

I am grateful to the many friends and colleagues who have offered valuable input over the years, ranging from ideas for problems to lively discussions of solutions. I would like to thank Jacob Barandes, Joe Blitzstein, Nancy Chen, Carol Davis, Louis Deslauriers, Eric Dunn, Niell Elvin, Dan Eniceicu, Howard Georgi, Theresa Morin Hall, Brian Hall, Lev Kaplan, Alex Johnson, Abijith Krishnan, Matt McIrvin, Lenny Ng, Dave Patterson, Sharad Ramanathan, Mike Robinson, Nate Salwen, Aravi Samuel, Alexia Schulz, Bob Silverman, Steve Simon, Igor Smolyarenko, Joe Swingle, Corri Taylor, Carey Witkov, Eric Zaslow, Tanya Zelevinsky, and Keith Zengel. My memory has certainly faded over the past 20 years, so I have surely left out other people who contributed to the book. Please accept my apologies!

Despite careful editing, there is zero probability that this book is error free. If anything looks amiss, please check for typos, updates, additional material, etc., at the webpage: www.people.fas.harvard.edu/~djmorin/book.html. And please let me know if you discover something that isn't already posted. Suggestions are always welcome.

David Morin
Cambridge, MA

List of Problems

General

	Problem	Hint	Solution
1. Green-eyed dragons	2	18	24
2. Simpson's paradox	2	18	27
3. Verifying weights	2	18	30
4. Counterfeit coin	2	18	32
5. The game of Nim	3	19	36
6. Monochromatic triangle	3	19	41
7. AM-GM Inequality	3	19	44
8. Crawling ant	3	19	46

Geometry

	Problem	Hint	Solution
9. Apple core	4	19	48
10. Viewing the spokes	4	19	51
11. Painting a funnel	4	19	53
12. Tower of circles	5	19	54
13. Ladder envelope	5	19	58
14. Equal segments	5	19	60
15. Collinear points	5	19	63
16. Attracting bugs	5	19	64
17. Find the foci	6	20	68
18. Construct the center	6	20	73
19. Find the angles	6	20	76
20. Rectangle in a circle	6	20	81
21. Product of lengths	6	20	83
22. Mountain climber	7	20	84

Probability

	Problem	Hint	Solution
23. Passing the spaghetti	8	20	88
24. How many trains?	8	20	90
25. Flipping a coin	9	20	92
26. Trading envelopes	9	20	93
27. Waiting for an ace	9	20	95
28. Drunken walk	9	21	99
29. HTH and HTT	10	21	101
30. Staying ahead	10	21	108
31. Random walk	10	21	111
32. Standing in a line	10	21	117
33. Rolling the die	11	21	118
34. Strands of spaghetti	11	21	120
35. How much change?	11	21	121
36. Relatively prime numbers	11	21	122
37. The hotel problem	11	22	124
38. Decreasing numbers	11	22	125
39. Sum over 1	11	22	129
40. Convenient migraines	12	22	130
41. Letters in envelopes	12	22	132
42. Leftover dental floss	12	22	138
43. Comparing the numbers	12	22	141
44. Shifted intervals	13	22	145
45. Intervals between independent events	13	22	147
46. The prosecutor's fallacy	14	22	149
47. The game-show problem	14	22	152
48. A random game-show host	14	23	155
49. The birthday problem	14	23	157
50. The boy/girl problem	15	23	165
51. Boy/girl problem with general information	15	23	169

Foundational

	Problem	Hint	Solution
52. Stirling's formula	15	23	172
53. A handy formula	16	23	176
54. Exponential distribution	16	23	179
55. Poisson distribution	16	23	184
56. Gaussian approximation to the binomial dist.	17	23	188
57. Gaussian approximation to the Poisson dist.	17	23	192

Chapter 1

Problems

TO THE READER: This book is available as both a paperback and an eBook. I have made the first chapter (the problems) available on the web, but it is possible (based on past experience) that a pirated version of the complete book will eventually appear on the web. In the event that you are reading such a version, I have a request:

If you don't find this book useful (in which case you probably would have returned it, if you had bought it), or if you do find it useful but aren't able to afford it, then no worries; carry on. However, if you do find it useful and are able to afford the Kindle eBook (priced below $10), then please consider purchasing it (available on Amazon). If you don't already have the Kindle reading app for your computer, you can download it free from Amazon. I chose to self-publish this book so that I could keep the cost low. The resulting eBook price of under $10 is less than a movie and a bag of popcorn, with the added bonus that the book lasts for more than two hours and has zero calories (if used properly!).

– David Morin

As mentioned in the preface (which you should be sure to read before tackling the problems), the most important advice for using this book is:

> **Don't look at the solutions too soon!**

The problems in this book are in general extremely difficult and are designed to be brooded over for a significant amount of time. Most of them are classics that have been around for ages, and there are only so many such problems in the world. If you look at a solution too soon, the opportunity to solve the problem is gone, and it's never coming back. Don't waste it!

1.1 General

1. **Green-eyed dragons**

    You visit a remote desert island inhabited by one hundred very friendly dragons, all of whom have green eyes. They haven't seen a human for many centuries and are very excited about your visit. They show you around their island and tell you all about their dragon way of life (dragons can talk, of course).

    They seem to be quite normal, as far as dragons go, but then you find out something rather odd. They have a rule on the island that states that if a dragon ever finds out that he/she has green eyes, then at precisely midnight at the end of the day of this discovery, he/she must relinquish all dragon powers and transform into a long-tailed sparrow. However, there are no mirrors on the island, and the dragons never talk about eye color, so they have been living in blissful ignorance throughout the ages.

    Upon your departure, all the dragons get together to see you off, and in a tearful farewell you thank them for being such hospitable dragons. You then decide to tell them something that they all already know (for each can see the colors of the eyes of all the other dragons): You tell them all that at least one of them has green eyes. Then you leave, not thinking of the consequences (if any). Assuming that the dragons are (of course) infallibly logical, what happens? If something interesting does happen, what exactly is the new information you gave the dragons?

2. **Simpson's paradox**

    During the baseball season in a particular year, player A has a higher batting average than player B. In the following year, A again has a higher average than B. But to your great surprise when you calculate the batting averages over the combined span of the two years, you find that A's average is *lower* than B's! Explain, by giving a concrete example, how this is possible.

3. **Verifying weights**

    (a) You have a balance scale and wish to verify the weight of an item that can take on any integral value from 1 to 121. What is the minimum number of fixed weights (with known values) you need, in order to cover all 121 possibilities? What are the weights? ("Verify" here means making the scale be balanced, so that you can be certain of the given item's weight.)

    (b) Using n wisely-chosen fixed weights, what is the largest integer W for which you can verify all the integral weights less than or equal to W? What fixed weights should you choose?

4. **Counterfeit coin**

    (a) You are given twelve coins, eleven of which have the same weight, and one of which has a weight different from the others (either heavier or lighter, you do not know). You have a balance scale. What is the minimum number of weighings required in order to guarantee that you can determine which coin has the different weight, and also whether it is heavier or lighter than the rest?

1.1. General

(b) You are given N coins, $N - 1$ of which have the same weight, and one of which has a weight different from the others (either heavier or lighter, you do not know). You are allowed W weighings on a balance scale. What is the maximum value of N, as a function of W, for which you are guaranteed to be able to determine which coin has the different weight, and also whether it is heavy or light?

5. **The game of Nim**

    Determine the best strategy for each player in the following two-player game. There are three piles, each of which contains some number of coins. Players alternate turns, each turn consisting of removing any (non-zero) number of coins from a single pile. The player who removes the last coin(s) wins.

6. **Monochromatic triangle**

    (a) Seventeen points, no three of which are collinear, are connected by all the possible lines between them ($\binom{17}{2} = 136$, in fact). Each line is colored either red, green, or blue (your choice for each line). Prove that within the resulting network of lines, there is at least one triangle all of whose sides are the same color.

    (b) Let $\lceil a \rceil$ denote the smallest integer greater than or equal to a. Let $\lceil n!e \rceil$ points, no three of which are collinear, be connected by all the possible lines between them. Each line is colored one of n colors (your choice for each). Prove that within the resulting network of lines, there is at least one triangle all of whose sides are the same color.

7. **AM-GM inequality**

    Prove the Arithmetic-Mean Geometric-Mean inequality:

    $$\frac{x_1 + x_2 + \cdots + x_n}{n} \geq \sqrt[n]{x_1 x_2 \cdots x_n}, \tag{1.1}$$

    where the x_i are non-negative real numbers. As a hint, the inequality can be proved by induction in the following way. Let I_n represent the above inequality. Show that I_2 is true, and then show that I_n implies I_{2n}, and then finally show that I_n implies I_{n-1}.

8. **Crawling ant**

    A rubber band with initial length L has one end attached to a wall. At $t = 0$, the other end is pulled away from the wall at constant speed V. (Assume that the rubber band stretches uniformly.) At the same time, an ant located at the end not attached to the wall begins to crawl toward the wall, with constant speed u relative to the band. Will the ant reach the wall? If so, how much time will it take?

1.2 Geometry

9. **Apple core**

    You find an apple core of height h. What volume of apple was eaten? (In this problem, an apple is a perfect sphere, and the height of the core is the height of the cylindrical part of its boundary.)

10. **Viewing the spokes**

    A wheel with radial spokes rolls (without slipping) on the ground. From off to the side, a stationary camera takes a picture of the wheel. If the exposure time is non-negligible, the spokes will in general appear blurred. At what locations in the picture do the spokes *not* appear blurred?

11. **Painting a funnel**

    Consider the curve $y = 1/x$, from $x = 1$ to $x = \infty$. Rotate this curve around the x-axis to create a funnel-like surface of revolution, as shown in Fig. 1.1.

    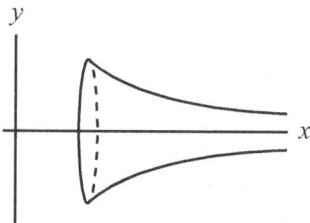

    Figure 1.1

    By slicing up the funnel into disks with radii $r = 1/x$ and thickness dx (and hence volume $(\pi r^2)\,dx$) stacked side by side, we see that the volume of the funnel is

    $$V = \int_1^\infty \frac{\pi}{x^2}\,dx = -\frac{\pi}{x}\bigg|_1^\infty = \pi, \qquad (1.2)$$

    which is finite. The surface area, however, involves the circumferential area of the disks, which is $(2\pi r)\,dx$ multiplied by a $\sqrt{1 + y'^2}$ factor accounting for the tilt of the area. The surface area of the funnel is therefore

    $$A = \int_1^\infty \frac{2\pi\sqrt{1+y'^2}}{x}\,dx > \int_1^\infty \frac{2\pi}{x}\,dx, \qquad (1.3)$$

    which is infinite because the integral of $1/x$, which is $\ln x$, diverges. (So the square-root factor turns out to be irrelevant for the present purposes.) Since the volume is finite but the area is infinite, it therefore appears that you can fill up the funnel with paint but you can't paint it. However, we then have a problem, because filling up the funnel with paint implies that you can certainly paint the *inside* surface. But the inside surface is the same as the outside surface, because the funnel wall has no thickness. So we should be able to paint the outside surface too. What's going on here? Can you paint the funnel or not?

1.2. Geometry

12. **Tower of circles**

    Consider N circles stacked on top of each other inside an isosceles triangle, as shown in Fig. 1.2 for the case of $N = 4$. Let A_C be the sum of the areas of the N circles, and let A_T be the area of the triangle. In terms of N, what should the vertex angle α be so that the ratio A_C/A_T is maximized? Assume that N is large, and ignore terms in your answer that are of subleading order in N. (Eq. (1.5) in Problem 53 might be helpful.)

    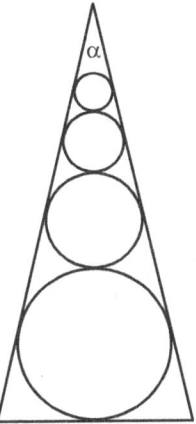

    Figure 1.2

13. **Ladder envelope**

    A ladder initially stands vertically against a wall. The bottom end is given a little sideways kick, causing the ladder to slide down. (The floor is slippery, so the ladder does in fact slide.) Assume that the bottom end is constrained to keep contact with the ground, and that the top end is constrained to keep contact with the wall. Describe the envelope of the ladder's positions.

14. **Equal segments**

    You are given a line segment, an (infinite) line parallel to it, and a straightedge. Show how to divide the segment into N equal segments, for any integer N. (With a straightedge, you are allowed only to draw straight lines and create intersections. You are not allowed to mark off distances on the straightedge.)

15. **Collinear points**

    You are given a finite number of points in space with the property that any line that contains two of the points contains three of them. Prove that all the points must lie on a common line.

16. **Attracting bugs**

    N bugs are initially located at the vertices of a regular N-gon whose sides have length ℓ. At a given moment, they all begin walking with equal speeds in the clockwise direction, directly toward the adjacent bug. They continue to walk directly toward the adjacent bug (whose position is continually changing, of course), until they finally all meet at the center of the original N-gon. What is the total distance each bug walks? How many times does each bug spiral around the center?

17. **Find the foci**

    Using a straightedge and compass, construct (1) the foci of a given ellipse, (2) the focus of a given parabola, and (3) the foci of a given hyperbola.

18. **Construct the center**

    Construct the center of a given circle, using only a compass. With a compass, you are allowed to mark points with the needle, and to draw arcs of circles (which may intersect at new points)

19. **Find the angles**

    Quadrilateral $ABCD$ in Fig. 1.3 has angles $\angle BAC = 80°$, $\angle CAD = 20°$, $\angle BDA = 50°$, and $\angle CDB = 50°$. Find angles $\angle BCA$ and $\angle CBD$.

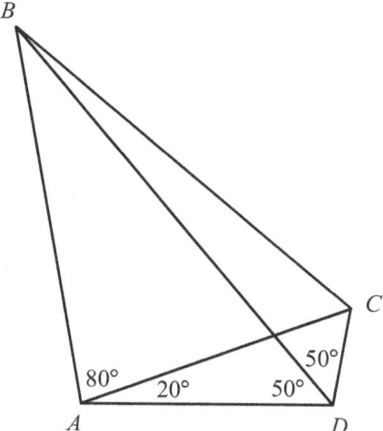

Figure 1.3

20. **Rectangle in a circle**

    Given a cyclic quadrilateral $ABCD$, draw the diagonals AC and BD. Prove that the centers of the inscribed circles of triangles ABC, BCD, CDA, and DAB are the vertices of a rectangle, as shown in Fig. 1.4.

21. **Product of lengths**

    Inscribe a regular N-gon in a circle of radius 1. Draw the $N - 1$ segments connecting a given vertex to the $N - 1$ other vertices. Show that the product of the lengths of these $N - 1$ segments equals N. Fig. 1.5 shows the case where $N = 10$; the product of the lengths of the nine segments is 10.

1.2. Geometry

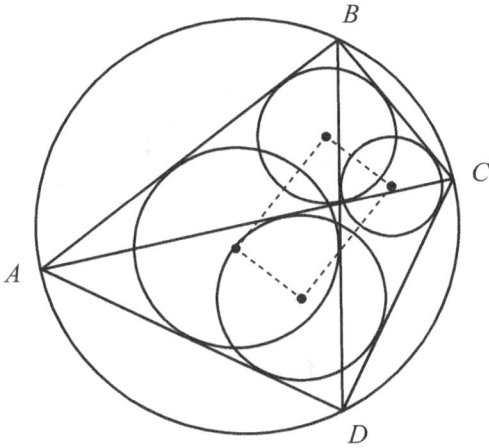

Figure 1.4

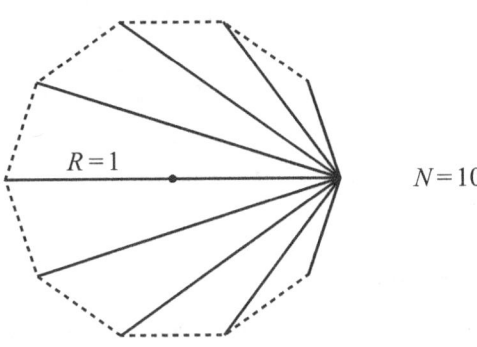

Figure 1.5

22. **Mountain climber**

    A mountain climber wishes to climb up a frictionless conical mountain. He wants to do this by throwing a lasso (a rope with a loop) over the top and climbing up along the rope. Assume that the climber is of negligible height, so that the rope lies along the mountain, as shown in Fig. 1.6.

    At the bottom of the mountain are two stores. One sells "cheap" lassos (made of a segment of rope tied to a loop of *fixed* length). The other sells "deluxe" lassos (made of one piece of rope with a loop of *variable* length; the loop's length may change without any friction of the rope with itself). See Fig. 1.7.

    When viewed from the side, the conical mountain has an angle α at its peak. For what angles α can the climber climb up along the mountain if he uses a cheap lasso? A deluxe lasso? (*Hint:* The answer in the cheap case isn't $\alpha < 90°$.)

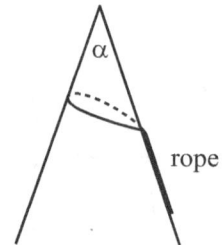

Figure 1.6

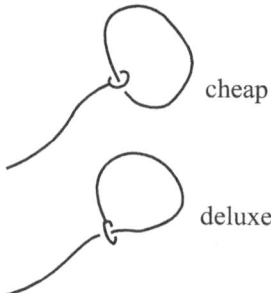

Figure 1.7

1.3 Probability

23. **Passing the spaghetti**

    At a dinner party, n people are seated around a table. A plate of spaghetti starts at the head of the table. The person sitting there takes some spaghetti and then passes the (very large) plate at random to their right or left, with a 50-50 chance of either direction. Henceforth each person receiving the plate takes some spaghetti and then passes the plate at random to their right or left. (Diners who have already received the plate can simply pass it on, without taking any more.) When all the diners have finally received their spaghetti, the plate stops being passed, and the eating begins.

    (a) What is the probability of being the last to be served, as a function of position (relative to the head) at the table of n people?

    (b) If this procedure is repeated over the course of many dinners, what is the average number of times the plate is passed?

24. **How many trains?**

    A train station consists of n parallel train tracks. The trains come at random times on each track, at equal average time intervals on each track. How many trains (including yours) will you see, on average, by the time the train on your track comes? The phrasing of this question is slightly ambiguous, so consider two possibilities:

1.3. Probability

(a) You arrive at the station at a random time and then count the trains until your train arrives. You repeat this process on many different days and take an average.

(b) You hang out at the station for a long time and count the trains (including yours) that arrive between each arrival of the train on your track. You then take an average.

25. **Flipping a coin**

(a) Consider the following game. You flip a coin until you get a tails. The number of dollars you win equals the number of coins you end up flipping. (So if you immediately get a tails, you win one dollar; if you get one heads before a tails, you win two dollars, etc.) What is the expectation value of your winnings?

(b) Play the same game, except now let the number of dollars you win be equal to 2^{n-1}, where n is the number of coins you end up flipping. What is the expectation value of your winnings now? Does your answer make sense?

26. **Trading envelopes**

(a) I give you an envelope containing a certain amount of money, and you open it. I then put into a second envelope either twice this amount or half this amount, with a 50-50 chance of each. You are given the opportunity to trade envelopes. Should you?

(b) I put two sealed envelopes on a table. One contains twice as much money as the other. You pick an envelope and open it. You are then given the opportunity to trade envelopes. Should you?

(c) If your answers to (a) and (b) are the same, explain why. If they are different, explain why.

27. **Waiting for an ace**

How many cards do you need to deal from a standard deck, on average, to get your first ace? A standard deck contains 52 cards, four of which are aces.

28. **Drunken walk**

A drunk performs a random walk along a street. (His steps are all of the same length and equally likely to be in either direction along the street.) At one end of the street is a river, and at the other end is a police station. If he gets to either of these ends, he remains there. He starts n steps from the river, and there are N total steps between the river and the police station.

(a) What is the probability that he ends up at the river? At the police station?

(b) What is the expected total number of steps he takes?

29. HTH and HTT

A coin is flipped repeatedly, and the resulting string of Heads and Tails is listed out. For example, the string might look like THHTHTTTHHT....

(a) Consider the two sequences of letters, HTH and HTT. Which sequence is more likely to occur first? Or are they equally likely to occur first?

(b) Let E_{HTH} be the expectation value for the number of flips needed to complete the first occurrence of HTH. For example, in the above string, the first HTH sequence is completed on the 5th flip. Likewise for E_{HTT}; in the above string, the first HTT sequence is completed on the 7th flip. Which of E_{HTH} and E_{HTT} is larger? Or are they equal?

(c) What are the values of E_{HTH} and E_{HTT}?

(d) In a large number of flips, how many times on average will each of HTH and HTT appear? (In the case of HTH sequences, a given H is allowed to count twice. For example, the string HTHTH contains two HTH sequences.)

30. Staying ahead

In a two-way election, candidate A receives a votes and candidate B receives b votes, with $a \geq b$. If the ballots are removed one at a time from the ballot box and a running total of the score is kept, what is the probability that at all times A's sub-total is greater than or equal to B's sub-total?

31. Random walk

Consider a random walk in one dimension. Each step has unit length, with equal probabilities of being rightward or leftward.

(a) What is the probability, p_{2n}, of returning to the origin (not necessarily for the first time) on the $(2n)$th step? (If you are at the origin, the number of steps must be even, of course.)

(b) What is the probability, f_{2n}, of returning to the origin for the first time on the $(2n)$th step? The technique used in Problem 30 is helpful here.

(c) Show that $f_{2n} = p_{2n-2} - p_{2n}$. A quick corollary is that $f_2 + f_4 + f_6 + \cdots = 1$, which means that you are guaranteed to eventually return to the origin in a 1-D random walk.

(d) Show that the probability, a_{2n}, of not returning to the origin at any time (even at the end) during a walk of length $2n$ equals p_{2n}. One method is to use the $f_{2n} = p_{2n-2} - p_{2n}$ result, but try to also think of a method that uses the technique from Problem 30.

32. Standing in a line

N people are standing in random order in a line, facing forward down the line. How many of them, on average, are able to say, "I am taller than everyone in front of me."?[1]

---

[1] There is a semantics issue with the first person in the line (who has no one in front of them). But let's declare that they are able to make the statement. Equivalently, we can work instead with the statement, "There is no one in front of me who is taller than I am." (Assume that no two people have exactly the same height.)

1.3. Probability

33. **Rolling the die**

    Two players alternately roll an N-sided die. The player who fails to improve upon the previous roll loses. What is the probability that the first player wins?

34. **Strands of spaghetti**

    A bowl contains N spaghetti noodles. You reach into the bowl and grab two free ends at random and attach them. You do this N times until there are no free ends left. On average, how many loops are formed by this process?

35. **How much change?**

    You are out shopping one day with $\$N$, and you find an item whose price takes on a random value between $\$0$ and $\$N$. (You may assume that $\$N$ is large compared with a penny, so that the distribution of prices is essentially continuous.) You buy as many of these items as you can with your $\$N$. What is the expectation value of the money you have left over?

36. **Relatively prime numbers**

    What is the probability that two randomly chosen positive integers are relatively prime (that is, they have no common factor, aside from 1)?

37. **The hotel problem**

    You are driving down a one-way road and pass a strip of a large number, N, of hotels. These all have different rates, arranged randomly. You want to maximize your chance of choosing the cheapest hotel, but you can't return to one you've passed up. Assume that your only goal is to obtain the cheapest hotel (the second cheapest is of no more value to you than the most expensive). If your strategy is to proceed past a certain fraction, x, of the hotels and then pick the next one that is cheaper than all the ones you've seen so far, what should x be? What, then, is your probability of success? Assume that N is very large, and ignore terms in your answer that are of subleading order in N.

38. **Decreasing numbers**

    Pick a random number (evenly distributed) between 0 and 1. Continue picking random numbers as long as they keep decreasing; stop picking when you obtain a number that is greater than the previous one you picked. What is the expected number of numbers you pick?

39. **Sum over 1**

    (a) You are given a random number (evenly distributed) between 0 and 1. To this, you add a second such random number. Keep adding numbers until the sum exceeds 1, and then stop. How many numbers, on average, will you need?

    (b) When the sum finally exceeds 1 and the game stops, what is the expectation value for the sum?

40. Convenient migraines

A probability course at a particular college has two exam days and 18 lecture days. At the end of the semester, the teacher notes that a certain student claimed to have a migraine headache on both of the exam days, thereby asking for the exams to be postponed. The teacher also notes that this student never had a headache during any of the 18 lectures.

The teacher is understandably suspicious of this coincidence, so she tries to calculate the probability P of having a headache on, and only on, the two exam days. But she quickly realizes, of course, that it is impossible to calculate P without knowing the probability p of a migraine occurring on a given day (which we'll assume is completely random and not based on real-life effects such as stress, etc.). Nevertheless, she realizes that it *is* possible to calculate the *maximum possible* value of P. And since any argument of plausibility on the student's part can use at most this best-case result, the teacher ends up coming to a fairly certain conclusion, as you will too.

(a) Let P be the probability of having a migraine on, and only on, the days of exams. If there are a exams and b lectures, what probability p of having a migraine on a given day leads to the maximum possible P?

(b) What is this maximum probability P_{max} in terms of a and b? What is P_{max} if $a = 2$ and $b = 18$?

(c) In the approximation where a is much smaller than b, and where b is assumed to be given, how does P_{max} depend on a?

41. Letters in envelopes

You are given N addressed letters and N addressed envelopes. If you randomly put one letter in each envelope, what is the probability that no letter ends up in the correct envelope?

42. Leftover dental floss

Two rolls of dental floss initially have equal lengths, L. Each day, a person chooses one of the rolls at random and cuts off a fixed small length, d. This continues until one of the rolls runs out of floss. How much floss, on average, is left on the other roll at this time? Assume that $N \equiv L/d$ is large, and ignore terms in your answer that are of subleading order in N. You will need to use a result from Problem 56.

43. Comparing the numbers

The integers 1 through N are put in a hat. (Technically, any set of N distinct numbers would work just as well.) You and $N - 1$ other people each pick a number. You then compare your number with the others, one at a time, until you find one that is smaller than yours. This procedure is repeated many times. How many numbers, on average, will you need to check in order to find one that is smaller than yours? Consider two cases:

(a) You ask the other people randomly. That is, at all times you have equal probabilities of asking each person. This could be arranged, for example,

1.3. Probability

by demanding that you have a very bad memory, so that you might ask a given person more than once.

(b) You have a good memory. In other words, you don't ask a given person more than once.

Ignore the scenarios where you have the number 1, because otherwise in part (a) the average would be infinite, and in part (b) you would always end up checking $N - 1$ numbers and never finding a smaller one.

44. Shifted intervals

Let $\epsilon \equiv 1/N$, where N is large. Choose a number at random between 0 and 1. Choose a second number between ϵ and $1 + \epsilon$. Choose a third number between 2ϵ and $1 + 2\epsilon$. Continue this process, until you choose an Nth number between $1 - \epsilon$ and $2 - \epsilon$. What is the probability that the first number you choose is the smallest of all the numbers? Assume that N is very large, and make suitable approximations.

45. Intervals between independent events

Consider a repeating event that happens completely randomly in time. Such a process can be characterized by the probability per unit time (call it p) of an event happening. The definition of p is that the probability of an event happening in an infinitesimal[2] time dt equals $p\,dt$.

From Problem 54 (you should solve that problem before this one), we know that starting at any particular time (not necessarily the time of an event), the probability that the next event happens between t and $t + dt$ later equals $pe^{-pt}\,dt$. (You can quickly show that the integral of this probability equals 1, as it must.) We're using p here in place of the λ from Problem 54.

(a) Using the $pe^{-pt}\,dt$ probability, show that starting at any particular time (not necessarily the time of an event), the average waiting time to the next event equals $1/p$. Explain why this is also the average time between events.

(b) Pick a random point in time, and look at the length of the time interval (between successive events) that it belongs to. Explain, using the above results, why the average length of this interval is $2/p$, and not $1/p$.

(c) We have found that the average time between events is $1/p$, and also that the average length of the interval surrounding a randomly chosen point in time is $2/p$. Someone might think that these two results should be the same. Explain intuitively why they are not.

(d) By correctly incorporating the probability distribution $pe^{-pt}\,dt$ mentioned above, show mathematically why $2/p$ is the correct result for the average length of the interval surrounding a randomly chosen point in time.

---

[2]The probability of an event happening in a *non*infinitesimal time t is *not* equal to pt. If t is large enough, then pt is larger than 1, so it certainly can't represent a probability. pt is the *average number* of events that occur in time t. But this doesn't equal the probability that an event occurs, because there can be double, triple, etc. events in the time t. We don't have to worry about multiple events if dt is infinitesimal.

46. The prosecutor's fallacy

Consider the following scenario. Detectives in a city, say, Boston (whose population we will assume to be one million), are working on a crime and have put together a description of the perpetrator, based on things such as height, a tattoo, a limp, an earing, etc. Let's assume that only one person in 10,000 fits the description. On a routine patrol the next day, police officers see a person fitting the description. This person is arrested and brought to trial based solely on the fact that he fits the description.

During the trial, the prosecutor tells the jury that since only one person in 10,000 fits the description (a true statement), it is highly unlikely (far beyond a reasonable doubt) that an innocent person fits the description (again a true statement), and therefore it is highly unlikely that the defendant is innocent. If you were a member of the jury, would you cast a "guilty" vote? If yes, what is your level of confidence? If no, what is wrong with the prosecutor's reasoning?

47. The game-show problem

A game-show host offers you the choice of three doors. Behind one of these doors is the grand prize, and behind the other two are goats. The host (who knows what is behind each of the doors) announces that after you select a door (without opening it), he will open one of the other two doors and purposefully reveal a goat. You select a door. The host then opens one of the other doors and reveals the promised goat. He then offers you the chance to switch your choice to the remaining door. To maximize the probability of winning the grand prize, should you switch or not? Or does it not matter?

48. A random game-show host

Consider the following variation of Problem 47. A game-show host offers you the choice of three doors. Behind one of these doors is the grand prize, and behind the other two are goats. The host announces that after you select a door (without opening it), he will *randomly* open one of the other two doors. You select a door. The host then randomly opens one of the other doors, and the result happens to be a goat. He then offers you the chance to switch your choice to the remaining door. Should you switch or not? Or does it not matter?

49. The birthday problem

(a) How many people need to be in a room in order for there to be a greater than 1/2 probability that at least two of them have the same birthday? By "same birthday" we mean the same day of the year; the year may differ. Ignore leap years.

(b) Assume that there is a large number N of days in a year. How many people are now necessary for the odds to favor a common birthday? Equivalently, assuming a normal 365-day year, how many people are required for there to be a greater than 1/2 probability that at least two of them were born in the same hour on the same date? Or in the same minute of the same hour on the same date? Neglect terms in your answer that are of subleading order in N.

1.4. Foundational

50. The boy/girl problem

The classic "boy/girl" problem can be stated in many different ways, with answers that may or may not be the same. Three different formulations are presented below, and a fourth is given in Problem 51. Assume in all of them that any process involved in the scenario is completely random. That is, assume that any child is equally likely to be a boy or a girl (even though this isn't quite true in real life), and assume that there is nothing special about the person you're talking with, and assume that there are no correlations between children (as there are with identical twins), and so on.

(a) You bump into a random person on the street who says, "I have two children. At least one of them is a boy." What is the probability that the other child is also a boy?

(b) You bump into a random person on the street who says, "I have two children. The older one is a boy." What is the probability that the other child is also a boy?

(c) You bump into a random person on the street who says, "I have two children, one of whom is this boy standing next to me." What is the probability that the other child is also a boy?

51. Boy/girl problem with general information

This problem is an extension of the preceding problem. You should study that one thoroughly before tackling this one. As in the original versions of the problem, assume that all processes are completely random. The new variation is the following:

You bump into a random person on the street who says, "I have two children. At least one of them is a boy whose birthday is in the summer." What is the probability that the other child is also a boy?

What if the clause is changed to "a boy whose birthday is on August 11th"? Or, "a boy who was born during a particular minute on August 11th"? Or more generally, "a boy who has a particular characteristic that occurs with probability p"?

1.4 Foundational

52. Stirling's formula

(a) Using $N! = \int_0^\infty x^N e^{-x} dx$ (which you can prove by induction), derive Stirling's formula,
$$N! \approx N^N e^{-N} \sqrt{2\pi N}. \tag{1.4}$$

Hint: Write $x^N e^{-x}$ as $e^{N \ln x - x} \equiv e^{f(x)}$, and then expand $f(x)$ in a Taylor series about its maximum, which you can show occurs at $x = N$. (See the appendix for a review of Taylor series.)

(b) Find also the first-order (in $1/N$) correction to Stirling's formula. (This calculation is a bit tedious.)

53. A handy formula

Expressions of the form $(1 + a)^n$ come up often in mathematics, especially in probability. Show that for small a,

$$(1 + a)^n \approx e^{na}. \tag{1.5}$$

Under what condition is this expression valid? Show that a more accurate approximation to $(1 + a)^n$ is

$$(1 + a)^n \approx e^{na} e^{-na^2/2}. \tag{1.6}$$

Under what condition is this expression valid? How should the righthand side be modified to make it even more accurate? There are various ways to answer these questions, but the cleanest way is to integrate both sides of the formula for the sum of an infinite geometric series,

$$1 - a + a^2 - a^3 + a^4 - \cdots = \frac{1}{1 + a}. \tag{1.7}$$

54. Exponential distribution

Consider a repeating event that happens completely randomly in time. By "completely randomly" we mean that there is a uniform probability that an event happens at any given instant (or more precisely, in any small time interval of a given length), independent of what has already happened. That is, the process has no "memory." Let the average time between events be τ. Equivalently, let $\lambda = 1/\tau$ be the average rate at which the events occur (the number per second, or whatever unit of time is being used).

The task of this problem is to derive the probability distribution $\rho(t)$ for the waiting time until the next event occurs is, that is, to determine the function $\rho(t)$ for which $\rho(t)\,dt$ is the probability that the next event occurs at a time between t and $t + dt$ (with $t = 0$ being when you start waiting). Show that $\rho(t)$ is given by

$$\rho(t) = \lambda e^{-\lambda t}. \tag{1.8}$$

This is (naturally) called the exponential distribution. You will need to use a result from Problem 53.

55. Poisson distribution

As with the exponential distribution in Problem 54, consider a repeating event that happens completely randomly in time. Show that the probability distribution for the number k of events that occur during a given time interval takes the form of the Poisson distribution,

$$P(k) = \frac{a^k e^{-a}}{k!}, \tag{1.9}$$

where a is the expected (average) number of events in the given interval. You will need to use a result from Problem 53. Note that whereas the exponential distribution deals with the *waiting time* until the next event, the Poisson distribution deals with the *number of events* in a given time (or space, etc.) interval.

1.4. Foundational

56. Gaussian approximation to the binomial distribution

If n coins are flipped, the probability of obtaining k Heads is given by the binomial distribution, $P(k) = \binom{n}{k}/2^n$. This follows from the fact that there are 2^n equally likely outcomes for the string of n flips, and $\binom{n}{k}$ of these outcomes have k Heads.

To make the math more tractable in this problem, let's replace n with $2n$, and k with $n + x$. So we're flipping $2n$ coins, and we want to get x Heads relative to the average (which is n). The probability is then

$$P_B(x) = \frac{1}{2^{2n}} \binom{2n}{n+x} \qquad \text{(for } 2n \text{ coin flips)} \qquad (1.10)$$

where the subscript B is for binomial. Show that if n is large, $P_B(x)$ takes approximately the Gaussian form,

$$P_B(x) \approx \frac{e^{-x^2/n}}{\sqrt{\pi n}} \equiv P_G(x) \qquad \text{(for } 2n \text{ coin flips)} \qquad (1.11)$$

where the subscript G is for Gaussian. You will need to use the results from Problems 52 and 53.

57. Gaussian approximation to the Poisson distribution

Show that in the limit of large a, the Poisson distribution from Problem 55,

$$P_P(k) = \frac{a^k e^{-a}}{k!}, \qquad (1.12)$$

takes approximately the Gaussian form,

$$P_P(x) \approx \frac{e^{-x^2/2a}}{\sqrt{2\pi a}}, \qquad (1.13)$$

where $x \equiv k - a$ is the deviation from the average, a. You will need to use the results from Problems 52 and 53.

Chapter 2

Hints

Most of the hints given here are fairly minimal, so you shouldn't expect a problem to be easy after reading the hint. If you need some further help, be sure to wait a while (at least a month) before looking at the solution. There's no hurry; just move on to another problem and occasionally come back to the original one. When you do eventually look at the solution, read only one line at a time, until you've read just enough to get a hint. Then set the book aside and try to work things out. That way, you'll still be able to (mostly) solve the problem on your own. Repeat as necessary, with a week between peeks at the solution.

1. GREEN-EYED DRAGONS: As with any problem involving a general number N, it is usually a good idea to first solve the problem for small values of N. For this problem, $N = 1$ is easy, $N = 2$ requires a little thought, and if you can solve $N = 3$, you've figured out the key point for inductively generalizing to arbitrary N.

2. SIMPSON'S PARADOX: The number of "at bats" doesn't need to be the same for the two players, nor the same in the two years. Try making them vastly different.

3. VERIFYING WEIGHTS: It's actually easier to do part (b) first, by using the standard strategy of solving a problem for small values of n. The $n = 2$ and $n = 3$ cases will allow you to see the pattern. To give a general proof, think about how many ways a given fixed weight can be used on the scale. This will give you an upper bound on the number of weights that can be verified with n fixed weights.

4. COUNTERFEIT COIN: Working with smaller numbers isn't terribly helpful in this problem, but you might want to first solve the case of three coins and two weighings. In general, there are three possible outcomes to each weighing: left side heavier, right side heavier, or the two sides equal. To gain as much information as possible from each weighing, all three possibilities should be realizable. For part (b), this three-possibilities fact provides a way of obtaining an upper bound (which might not be achievable) on the number N of coins that can be solved with W weighings. It may be helpful to consider the setup where you have an additional known good coin at your disposal.

5. THE GAME OF NIM: To get a feel for the problem, you should start by working with small numbers. It is helpful to make a list of triplets of numbers that guarantee you lose if you encounter them (assuming that your opponent is playing optimally). You can make a table with the two axes being two numbers in the triplet, and the entry in the table being the third. This table might lead you to believe that powers of 2 are relevant, so the main hint (stop reading now if you don't want this hint yet) is to write the numbers for a losing triplet in base 2, see how they relate, and then make a conjecture. Proving the conjecture is another matter.

6. MONOCHROMATIC TRIANGLE: First solve the trivial problem of three points and one color, and then six points and two colors. In general, to go from one case to the next, isolate one point and use induction along with the pigeonhole principle.

7. AM-GM INEQUALITY: To demonstrate the $n = 2$ case, use the fact that $(a-b)^2 \geq 0$.

8. CRAWLING ANT: It is perhaps easiest to think in terms of the fraction of the way (from the moving end to the wall) the ant is, as a function of t. The question is then whether this fraction ever equals 1. How does the fraction change in a small time dt?

9. APPLE CORE: Find the cross sectional area (produced by a horizontal plane) of the eaten part. Do you know of another object with the same cross sectional area?

10. VIEWING THE SPOKES: A spoke doesn't appear blurred at a given point in the picture if the point lies along the same spoke (perhaps at different points on the spoke) throughout the duration of the camera's exposure.

11. PAINTING A FUNNEL: Think about units/dimensions.

12. TOWER OF CIRCLES: Let r be the ratio of the radii of two successive circles. You can find all relevant quantities in terms of r. You will need to use Eq. (1.5) in Problem 53 when dealing with large N.

13. LADDER ENVELOPE: The problem is equivalent to finding the locus of intersections between adjacent positions of the ladder. Let the two adjacent positions make angles $\theta + d\theta$ and θ with the floor. Find the coordinates of the intersection in terms of θ, and then convert them to an equation involving x and y.

14. EQUAL SEGMENTS: Pick an arbitrary point P on the side of AB opposite to the given infinite line L. If you draw every possible line you can draw in 15 seconds, you'll undoubtedly divide AB into two equal segments, whether you know it or not. You just need to prove why. Drawing a few more lines will inevitably give you three equal segments. Then show inductively how to go from N to $N + 1$.

15. COLLINEAR POINTS: Assume that the points don't all lie on a common line. Consider the shortest (nonzero) distance from a point to a line, and generate a contradiction.

16. ATTRACTING BUGS: The setup always remains an N-gon; it simply shrinks in size. Find the rate at which the distance between adjacent bugs decreases.

17. FIND THE FOCI: An ellipse is a stretched circle. Find a construction for a circle that isn't ruined by the stretching. The parabola and hyperbola cases are similar.

18. CONSTRUCT THE CENTER: Construct a rhombus (with an arbitrary side length ℓ) with three vertices lying on the circle. The length of one of the diagonals will involve ℓ and the radius of the circle, R. You now just need to perform another construction to get rid of the arbitrary length ℓ.

19. FIND THE ANGLES: Angle chasing won't get the job done here. Try drawing some new lines, and look for similar triangles. In particular (but stop reading now if you don't want the following hint), draw the angle bisectors of triangle ACD.

20. RECTANGLE IN A CIRCLE: Let the incenters of triangles ADB and ADC be X and Y, respectively. Draw the diagonals of quadrilateral $ADYX$ and find whatever angles you can in terms of arcs of the circle. Then look for similar triangles.

21. PRODUCT OF LENGTHS: The cleanest way to solve this problem is to use complex numbers. Put the circle in the complex plane, with its center at the origin and the given vertex at $(1, 0)$. If $a \equiv e^{2\pi i/N}$, then the other vertices are located at points of the form a^n (the Nth roots of 1). Consider the function $F(z) \equiv z^N - 1$, and think about ways to factor it.

22. MOUNTAIN CLIMBER: Imagine cutting a paper cone along a line emanating from the top, and then rolling it onto a plane. This can be done without crumpling the paper, because a cone is "flat." What shape must the path of the lasso take on the rolled-out cone? This puts a constraint on the cheap-lasso case. For the deluxe lasso, you don't want it to be possible for the mountain climber to fall by means of the lasso's loop changing its length.

23. PASSING THE SPAGHETTI: Solve the problem for $n = 2, 3, 4$, and then make a conjecture. Think about the things that need to happen in order for a given diner to be the last to be served. To find the average number of times the plate is passed, use an inductive argument.

24. HOW MANY TRAINS?: Find the probability that the kth train is yours, while the first $k - 1$ trains are not, and then use this to find the expected number of trains you see, by the time yours arrives. You will encounter a sum that you will want to write as the sum of an infinite number of different geometric series.

25. FLIPPING A COIN: In part (a), you will encounter a sum that you will want to write as the sum of an infinite number of different geometric series. If you encounter a strange result in part (b), think about what the exact definition of "expectation value" is.

26. TRADING ENVELOPES: For part (b), try to explicitly construct a setup where, once you pick an envelope, the other envelope has a 50-50 chance of containing twice or half the amount in your envelope.

27. WAITING FOR AN ACE: To be general, let the deck consist of N cards, n of which are aces. By playing around with some small values of N and n, you can guess the form of the answer. Then you can prove it by induction.

28. DRUNKEN WALK: A recursion relation is the standard first step in solving both parts of this problem.

29. HTH AND HTT: Start with shorter sequences: Find the expected waiting time to get an H, and also to get an HT. This will lead you to the waiting times for HTH and HTT, via a recursion formula.

30. STAYING AHEAD: Consider a two-dimensional lattice in which a vote for A is signified by a unit step in the positive x-direction, and a vote for B is signified by a unit step in the positive y-direction. The counting of the votes corresponds to a path from the origin to the point (a, b), with $a \geq b$. (How do the probabilities of the different paths compare?) We are concerned with paths that reach (a, b) without passing through the $y > x$ region (which means that the first step must be to $(1, 0)$). The paths starting at $(1, 0)$ that *do* enter the $y > x$ region must touch the line $y = x + 1$. The number of such paths can be found via a clever technique involving a reflection across this line; reflect the portion of the path between $(1, 0)$ and the first point of contact with the line $y = x + 1$. You can convince yourself why this method of counting works.

31. RANDOM WALK: For part (b), you will want to associate the 1-D random walk with a walk in the 2-D x-y plane, and then use the reflection technique from Problem 30.

32. STANDING IN A LINE: An induction argument (adding a new person at the back of the line) is probably the quickest way to solve this problem.

33. ROLLING THE DIE: If you work things out for a few small values of N, and if you instead look at the probability that the first player *loses*, you will see a pattern and can then make a conjecture. To prove it, it is perhaps best to derive the (more general) formula for the probability L_r that a player loses, given that a roll of r has just occurred. To do this, write down all the ways that the player can win (which happens with probability $1 - L_r$). You can eventually set $r = 0$.

34. STRANDS OF SPAGHETTI: For each chosen pair of ends, imagine picking them in succession instead of grabbing them simultaneously; this doesn't affect the process. At each stage, there are only two fundamentally different possibilities for what can happen, and a certain quantity doesn't depend on which of these occurs. Note that the length of a strand is irrelevant in this problem.

35. HOW MUCH CHANGE?: If the item costs between $N/2$ and N dollars, then you can buy only one item, and you have a certain average amount of money left over. Continue downward in price, with the analogous intervals that allow you to buy given numbers of items. It may be helpful to make a plot of your change vs. the price.

36. RELATIVELY PRIME NUMBERS: First write down the probability that two integers don't have a given prime p as a common factor. Then use the equality $1/(1-x) = 1 + x + x^2 + x^3 + \cdots$, along with the Unique Factorization Theorem.

37. THE HOTEL PROBLEM: It is helpful to organize the different cases according to what the highest-ranking hotel is (in order of cheapness) in the first fraction x. For each case, you can calculate the probability of success in terms of x.

38. DECREASING NUMBERS: Find the probability that a list of n random numbers is in decreasing order. Likewise for $n + 1$, and then put the results together.

39. SUM OVER 1: A helpful sub-step is to show that given n random numbers between 0 and 1, the probability $P_n(1)$ that their sum does not exceed 1 equals $1/n!$.

40. CONVENIENT MIGRAINES: For part (c) you will need to use Eq. (1.5) in Problem 53.

41. LETTERS IN ENVELOPES: Let B_N denote the number of "bad" arrangements where none of the N letters end up in the correct envelope. Try to find a recursion relation for B_N. If you generate a bad arrangement by introducing an $(N + 1)$th letter and envelope to a given arrangement of N letters and envelopes, there are only a couple possibilities for what was going on with the N letters and envelopes.

42. LEFTOVER DENTAL FLOSS: The act of choosing a roll can be mapped onto a random walk in a 2-D plane. You want to find the probability that the process ends at the point (N, n), in which case a length $(N - n)d$ is left on a roll. To generate an approximate answer for large N, you will need to use the fact that a binomial coefficient can be approximated by a Gaussian function; see Problem 56.

43. COMPARING THE NUMBERS: If your number is n, find the expected number of numbers you need to check, in order to find one that is smaller than yours. (A good strategy is to use a recursion-type of reasoning, taking into account the possibilities for what can happen on your first check.) Then average over the n's.

44. SHIFTED INTERVALS: Discretize each of the intervals (with length 1) into sub-intervals with (tiny) length ϵ. Consider the different cases where the first number is in each sub-interval, and find the probability that the first number is the smallest, for each case. To make an approximation for large N, take the log of the probabilities and use the Taylor approximation for $\ln(1 - x)$.

45. INTERVALS BETWEEN INDEPENDENT EVENTS: The key point in this problem is that there is a difference between the average length of an interval that is randomly chosen from a long sequence of intervals (all of which are equally likely to be chosen) and the average length of the interval that contains a randomly chosen point in time. Think about why these two averages should be different. Are all of the intervals equally likely in the latter case?

46. THE PROSECUTOR'S FALLACY: Break the prosecutor's reasoning down into a series of if/then statements, and determine the probability associated with each statement. Write each probability as a conditional probability $P(A|B)$, which is shorthand for "the probability of A, given B." A Venn-type diagram is helpful.

47. THE GAME-SHOW PROBLEM: The best hint here is to just play the game a number of times. Three standard playing cards, one red and two black, will do the trick. Even if you're positive of an answer you arrived at by theoretical means, you should still check it by playing the game!

48. A RANDOM GAME-SHOW HOST: As with Problem 47, you should just play the game a number of times. The randomness can be determined by a coin toss. You will end up throwing away some of the games (the ones where the host reveals the prize), because the condition of this problem is that he happens to reveal a goat.

49. THE BIRTHDAY PROBLEM: It's much easier to calculate the probability that there *isn't* a common birthday, and then subtract this from 1. Imagine plopping birthdays down on a calendar, one at a time. For part (b), take a log and use the Taylor approximation for $\ln(1 - x)$.

50. THE BOY/GIRL PROBLEM: Make a list of the various equally likely possibilities for the family's children, while taking into account only the "I have two children" information, and not yet the information about the boy. Then use the latter information to eliminate some of the possibilities.

51. BOY/GIRL PROBLEM WITH GENERAL INFORMATION: As in the preceding problem, make a list (or rather, a table here) of the various possibilities (perhaps involving differing probabilities), while taking into account only the "I have two children" information. Then use the new information to eliminate some of the possibilities.

52. STIRLING'S FORMULA: For part (a), the proof by induction can be done via integration by parts. You will then need to actually derive the Taylor series for $f(x)$ by taking some derivatives. And you will need to use the fact that $\int_{-\infty}^{\infty} e^{-x^2/b} dx = \sqrt{b\pi}$. For part (b), more terms in the Taylor series are necessary, along with values of integrals of the form $\int_{-\infty}^{\infty} x^{2n} e^{-ax^2} dx$, which can be obtained by differentiating $\int_{-\infty}^{\infty} e^{-ax^2} dx = \sqrt{\pi/a}$ with respect to a.

53. A HANDY FORMULA: After the suggested integration, exponentiate both sides of the equation, and you're almost there.

54. EXPONENTIAL DISTRIBUTION: First, convince yourself that if dt is very small, then $\lambda\, dt$ is the probability that an event occurs in a given interval with length dt. Divide the time t into many small intervals and calculate the probability that there is failure in every one of these intervals, along with success in a dt interval tacked on the end.

55. POISSON DISTRIBUTION: As in Problem 54, if dt is very small, then $\lambda\, dt$ is the probability that an event occurs in a given interval with length dt. Divide the time t into many small intervals and calculate the probability that k of them yield success; the result will involve a binomial coefficient. You will then need to make a number of approximations.

56. GAUSSIAN APPROXIMATION TO THE BINOMIAL DISTRIBUTION: Use Stirling's formula to rewrite the binomial coefficient, and simplify the result as much as you can. Then use the approximation from Eq. (1.6) in Problem 53. You will need to keep terms of order x^2 in the exponential.

57. GAUSSIAN APPROXIMATION TO THE POISSON DISTRIBUTION: Use Stirling's formula to rewrite $k!$, and simplify the result as much as you can. Then, as in Problem 56, use Eq. (1.6) and keep terms of order x^2.

Chapter 3

Solutions

1. **Green-eyed dragons**

    Let's start with a smaller number of dragons, N, instead of 100, to get a feel for the problem.

    If $N = 1$, and if you tell this dragon that at least one of the dragons has green eyes, then you are simply telling him that he has green eyes, so he must turn into a sparrow at midnight.

    If $N = 2$, let the dragons be labeled A and B. After your announcement that at least one of them has green eyes, B will think to himself, "If I do *not* have green eyes, then A can see that I don't, so A will conclude that she must have green eyes. She will therefore turn into a sparrow on the first midnight." Therefore, if A does not turn into a sparrow on the first midnight, then on the following day B will conclude that he himself must have green eyes, and so he will turn into a sparrow on the second midnight. The same thought process will occur for A, so they will both turn into sparrows on the second midnight.

    If $N = 3$, let the dragons be labeled A, B, and C. (Going from $N = 2$ to $N = 3$ contains the key logical step here. If we can solve the problem for $N = 3$, we'll be well on our way to solving the problem for general N.) After your announcement, C will think to himself, "If I do *not* have green eyes, then A and B can see that I don't. So I am irrelevant as far as they are concerned, which means that they can use the reasoning for the $N = 2$ situation, in which case they will both turn into sparrows on the second midnight." Therefore, if A and B do not turn into sparrows on the second midnight, then on the third day C will conclude that he himself must have green eyes, and so he will turn into a sparrow on the third midnight. The same thought process will occur for A and B, so they will all turn into sparrows on the third midnight. The pattern is now clear:

    Claim: *Consider N dragons, all of whom have green eyes. If you announce to all of them that at least one of them has green eyes, then they will all turn into sparrows on the Nth midnight.*

Proof: We will prove this by induction. We will assume that the result is true for N dragons, and then we will show that it is true for $N + 1$ dragons. We saw above that it is true for $N = 1, 2, 3$.

Consider $N + 1$ dragons, and pick one of them, labeled A. After your announcement, she will think to herself, "If I do *not* have green eyes, then the N other dragons can see that I don't. So I am irrelevant as far as they are concerned, which means that they can use the reasoning for the setup with N dragons, in which case they will all turn into sparrows on the Nth midnight." Therefore, if they do not all turn into sparrows on the Nth midnight, then on the $(N + 1)$th day A will conclude that she herself must have green eyes, and so she will turn into a sparrow on the $(N + 1)$th midnight. The same thought process will occur for the other N dragons, so they will all turn into sparrows on the $(N + 1)$th midnight. ∎

Therefore, in the given problem with 100 dragons, they will all turn into sparrows on the 100th midnight.

Although we've solved the problem, you may be troubled by the fact that your seemingly useless information did indeed have major consequences. How could this be, when all of the dragons already knew what you told them? Did you really give them new information? The answer is "yes." Let's see what this new information is.

Consider the $N = 1$ case. Here it is clear that you provided new information, because you essentially told the one dragon that he has green eyes. But for $N \geq 2$, the new information is more subtle.

Consider the $N = 2$ case. Prior to your announcement, A knows that B has green eyes, and B knows that A has green eyes. That is the extent of the knowledge, and the dragons can't conclude anything else from it. But after you tell them that at least one of them has green eyes, then A knows *two* things: He knows that B has green eyes, *and* he knows that B knows that there is at least one dragon with green eyes (because A knows that B heard your information). A had no way of knowing this fact before your announcement. B gains a similar second piece of information. This second piece of information is critical, as we saw above in the reasoning for the $N = 2$ case.

Consider the $N = 3$ case. Prior to your announcement, A knows that B has green eyes (and also that C has green eyes, but let's concentrate on B for now). And A also knows that B knows that there is at least one dragon with greens eyes, because A can see that B can see C. So the two bits of information in the $N = 2$ case above are already known before you speak. What new information is gained when you speak? Only after you speak is it true that A knows that B knows that C knows that there is at least one dragon with green eyes (because A knows that B knows that C heard your information, because all the dragons are standing right there). To be clear, before you speak, B certainly does know that C knows that there is at least one dragon with green eyes, because B can see that C can see A. (Note that A's green eyes are necessary here, because for all B knows, he (B) doesn't have green eyes.) However, A doesn't *know* this yet, because for all A knows, he (A) doesn't have green eyes.

The analogous result holds for a general number N. So there is no paradox here. Information *is* gained by your announcement. More information is added to the world than the information you gave.[1] And it turns out, as seen in the proof of the above claim, that the new information is enough to allow all the dragons to eventually figure out their eye color.

To sum up: Before you make your announcement, the following statement is valid in the case of N dragons: A_1 knows that A_2 knows that A_3 knows that... that A_{N-2} knows that A_{N-1} knows that there is at least one dragon with green eyes. (Note that this chain contains only $N - 1$ dragons.) This is true because A_{N-1} can see A_N; and A_{N-2} can see that A_{N-1} can see A_N; and so on, until lastly A_1 can see that A_2 can see that... that A_{N-1} can see A_N. The same result holds, of course, for any permutation of any group of $N - 1$ dragons. But only after you make your announcement is it true that the "A_1 knows that A_2 knows that A_3 knows that..." chain extends the final step to the Nth dragon. The truth of the complete chain relies critically on the fact that the Nth dragon heard your announcement (and that all the dragons know he heard it). So in the end, it turns out to be of great importance how far the "A knows that B knows that C knows that..." chain goes.

You might wonder if there is any additional logic we missed above that causes the dragons to turn into sparrows *before* the 100th midnight. As an exercise, you can show that this isn't the case. (Start with some small values of N.)

REMARKS:

1. If one of the dragons misses your farewell announcement (which is that at least one the 100 dragons on the island has green eyes), and if this absence is noted by all the other dragons, then they will all happily remain dragons throughout the ages. (You should verify this by starting with some small values of N.) In the statement of the problem, the seemingly innocuous word "all" in the phrase, "Upon your departure, *all* the dragons get together to see you off...," is critical to the logic.

2. If, on the other hand, all of the dragons are present, but one of them is daydreaming (as dragons tend to do) and doesn't hear your announcement, and if none of the other dragons notice this, then on the 100th midnight the other 99 dragons will turn into sparrows. The daydreamer will be left to wonder what happened. However, given that the transformation happened on the 100th instead of the 99th midnight, and given that dragons are infallibly logical, the daydreamer will undoubtedly suspect that he is playing a role in some sort of logic puzzle and furthermore that he has green eyes. But since there is no way to be certain, he will continue to live on in blissful dragon ignorance.

3. The original problem dealt with the case where every dragon hears your announcement and every dragon knows they all hear. We then looked at the

---

[1]For example, A knows that you made your announcement *while* stepping onto your boat *and* wearing a blue shirt. Or, more relevantly, A knows that you made your announcement *in front of all the other dragons*. In short, it's not just what you say. It's how (or more relevantly here, to whom) you say it.

case (in the first remark above) where one dragon misses the announcement and every dragon knows he missed it. We then looked at the case (in the second remark above) where one dragon misses the announcement but the other dragons don't notice this; the particular dragon *doesn't* hear the announcement, but the other dragons think he *does*. There is a fourth permutation we haven't dealt with yet: What if a particular dragon *does* hear the announcement, but the other dragons think he *doesn't*. For example, he is hiding in the bushes and listening in. (The dragons all know each other well, so his green eyes are still known to everyone even though they can't see him at this moment.) As an exercise, you can think about what happens in this case. ♣

2. Simpson's paradox

The two tables in Table 3.1 show an extreme scenario that gets to the heart of the matter. In the first year, player A has a small number of at-bats (6), while player B has a large number (600). In the second year, these numbers are reversed. You should examine these tables for a minute to see what's going on, before reading the next paragraph.

	First year	Second year
Player A	3/6 (.500)	150/600 (.250)
Player B	200/600 (.333)	1/6 (.167)

	Combined years
Player A	153/606 (.252)
Player B	201/606 (.332)

Table 3.1: Yearly and overall batting averages. The years with the large numbers of at-bats dominate the overall averages.

The main point to realize is that in the combined span of the two years, A's average is dominated by the .250 average coming from the large number of at-bats in the second year (yielding an overall average of .252, very close to .250), whereas B's average is dominated by the .333 average coming from the large number of at-bats in the first year (yielding an overall average of .332, very close to .333). B's .333 is lower than A's .500 in the first year, but that is irrelevant because A's very small number of at-bats that year hardly affects his overall average. Similarly, B's .167 is lower than A's .250 in the second year, but again, that is irrelevant because B's very small number of at-bats that year hardly affects his overall average. What matters is that B's .333 in the first year is higher than A's .250 in the second year. The large numbers of associated at-bats dominate the overall averages.

Fig. 3.1 shows a visual representation of the effect of the number of at-bats. The size of a data point in the figure gives a measure of the number of at-bats. So although B's average is lower than A's in each year, the large B data point in the first year is higher than the large A data point in the second year. These data points are what dominate the overall averages.

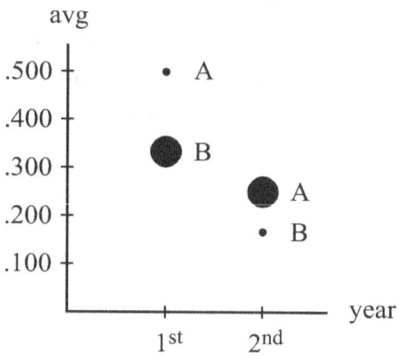

Figure 3.1

REMARKS:

1. To generate the paradox where B's overall average surprisingly ends up being higher than A's overall average, the higher of B's two yearly averages must be higher than the lower of A's two yearly averages. If this weren't the case (that is, if the large B data point in Fig. 3.1 were lower than the large A data point), then A's overall average would necessarily be higher than B's overall average (as you can verify). So the paradox wouldn't be realized.

2. To generate the paradox, we must also have a disparity in the number of at-bats. If all four of the yearly at-bats in the first of the tables in Table 3.1 were the same (or even just the same within each year, or just the same for each person), then A's overall average would necessarily be higher than B's overall average (as you can verify). The main point of the paradox is that when calculating the overall average for a given player, we can't just take the average of the two averages. A year with more at-bats influences the average more than a year with fewer at-bats, as we saw above.

   The paradox can certainly be explained with at-bats that don't have values as extreme as 6 and 600, but we chose these in order to make the effect as clear as possible. Also, we chose the total number of at-bats in the above example to be the same for A and B over the two years, but this of course isn't necessary.

3. The paradox can also be phrased in terms of averages on exams, for example: For 10th graders taking a particular test, boys have a higher average than girls. For 11th graders taking the same test, boys again have a higher average than girls. But for the 10th and 11th graders combined, girls have a higher average than boys. This can occur, for example, if 11th graders scored sufficiently higher than 10th graders in general, and if most 11th graders are girls while most 10th graders are boys. See Fig. 3.2.

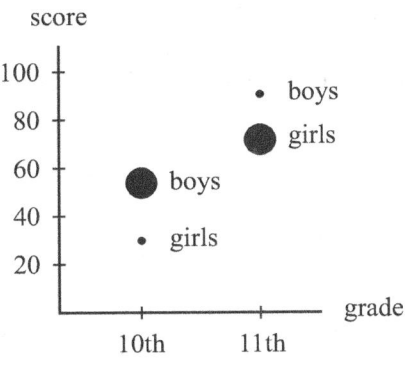

Figure 3.2

Another real-life example deals with college admissions rates. The paradox can arise when looking at male/female acceptance rates to individual departments, and then looking at the male/female acceptance rates to the college as a whole. (The departments are analogous to the different baseball years or to the different grades.)

4. One shouldn't get carried away with Simpson's paradox. There are plenty of scenarios where it doesn't apply, for example: In the 10th grade in a particular school, the percentage of students who are soccer players is larger than the percentage who are musicians. And in the 11th grade, the percentage of students who are soccer players is again larger than the percentage who are musicians. Can the overall percentage of students who are soccer players (in the combined grades) be smaller than the overall percentage who are musicians? (Think about this before reading further.)

The answer is a definite "No." One way to see why is to consider the *numbers* of soccer players and musicians, instead of the *percentages*. Since there are more soccer players than musicians in each grade, the total number (and hence percentage) of students who are soccer players must be larger than the total number (and hence percentage) who are musicians.

Another way to understand the "No" answer is to note that when calculating the percentages of students who are soccer players or musicians in a given grade, we are dividing the number of students of each type by the *same* denominator (namely, the total number of students in the grade). We therefore can't take advantage of the effect in the original baseball scenario above, where B's average was dominated by one year while A's was dominated by a different year due to the different numbers of at-bats in a given year. Instead of the data points in Fig. 3.1, the present setup might yield something like the data points in Fig. 3.3. The critical feature here is that the dots in each year have the *same* size. The dots for the 11th grade happen to be larger because we're arbitrarily assuming that there are more students in that grade. The total percentage of students who are soccer players in the two years is the weighted average of the two soccer dots (weighted by the size of the dots, or equivalently by the number of students in each grade). Likewise for the two music dots. The soccer weighted average is necessarily larger than the music weighted average. (This is fairly clear intuitively, but as an exercise you can prove it rigorously if you have your doubts.) ♣

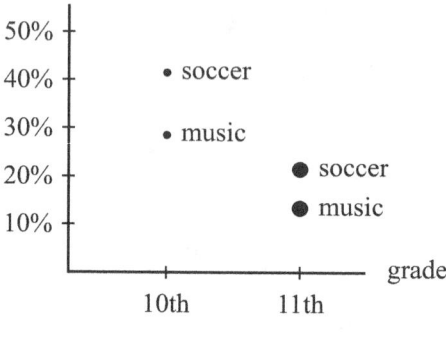

Figure 3.3

3. **Verifying weights**

   (a) Each fixed weight can be used in three different ways. It can be put on the left side of the scale, on the right side, or not used at all. Therefore, if we have n weights, they may be combined in 3^n ways. This is true because in adding the weights, there are three choices for the coefficient of each weight in the sum: a plus sign if it's on the left side of the scale, a minus sign if it's on the right, or a zero if it's not used at all.

   There are, however, duplicates among these 3^n combinations. For every positive number, there is its negative (where the left and right scales are reversed), which represents the same net weight on the balance scale. Since it is, in principle, possible for the number 0 not to be repeated in the 3^n combinations, an upper bound on the number of positive integer weights that can be verified with n fixed weights is $(3^n - 1)/2$. Therefore, to weigh all weights up to 121, we must have $n \geq 5$, since $121 = (3^5 - 1)/2$.

   We claim that five weights are in fact sufficient. We can show this by explicitly picking a set of five weights that get the job done. Let the first weight be 1 (to get the value 1). Then let the second weight be 3 (to get values up to 4; for example, 2 is obtained by putting the 3 on one side of the scale and the 1 on the other). Then let the third weight be 9 (to get values up to 13; for example, 5 is obtained by putting the 9 on one side and the 3 and 1 on the other). Then let the fourth weight be 27 (to get values up to 40, as you can check). And then let the fifth weight be 81 (to get values up to 121, as you can check). The fixed weights are therefore powers of 3.

   (b) In general, the n weights of $1, 3, 9, \ldots, 3^{n-1}$ may be used to verify all integral weights from 1 to $W_n \equiv (3^n - 1)/2$. This can be proved by induction on n, as follows.

   Assume that all weights up to $W_n = (3^n - 1)/2$ can be verified with n fixed weights. If the $(n + 1)$th fixed weight is chosen to be 3^n, then we can additionally verify all weights from $3^n - W_n$ (by putting the 3^n and W_n on opposite sides of the scale) to $3^n + W_n$ (by putting the 3^n and W_n on the same side of the scale). Using the above form of W_n, this range can be rewritten as $(3^n + 1)/2$ to $(3^{n+1} - 1)/2$. But this is exactly the range needed to be able to verify any weight up to $W_{n+1} = (3^{n+1} - 1)/2$, because the numbers up to $(3^n + 1)/2 - 1 = (3^n - 1)/2$ were already covered, by the inductive hypothesis.

Therefore, we have shown that if we can verify up to W_n by using n fixed weights, then we can also verify up to W_{n+1} by using $n + 1$ fixed weights. Since we can clearly verify up to $W_1 = 1$ by using one weight, the result holds for all n.

We have shown that it is possible to verify up to $W_n \equiv (3^n - 1)/2$ with n fixed weights. Additionally, we can't go any higher than this, because we know from the argument in part (a) that $W_n \equiv (3^n - 1)/2$ is an upper bound on the number of positive integer weights that can be verified with n fixed weights. Therefore, $W_n \equiv (3^n - 1)/2$ is the answer to the given question, with the fixed weights being powers of 3, running from $3^0 = 1$ up to 3^{n-1}.

REMARK: To rephrase the inductive argument in part (b), powers of 3 have the relevant property that $(1 + 3 + 3^2 + \cdots + 3^{n-1}) + 1 = 3^n - (1 + 3 + 3^2 + \cdots + 3^{n-1})$, as you can check. The importance of this equality can be seen by looking at, for example, $n = 4$. We see that 41, which is not quite obtainable with four weights (which yield up to $1 + 3 + 9 + 27 = 40$), can be obtained by taking the fifth weight, 81, and subtracting off the highest possible sum of four weights, namely 40, by putting this sum on the other side of the scale.

More generally, we can pretend that we don't already know that the fixed weights are powers of 3. If we label them as k_i, then the equality in the preceding paragraph takes the form of $(k_1 + k_2 + \cdots + k_n) + 1 = k_{n+1} - (k_1 + k_2 + \cdots + k_n)$. That is, 1 more than that largest number obtainable with n weights can be obtained by taking the $(n + 1)$th weight and subtracting off the highest possible sum of n weights. This relation can be rewritten as $k_{n+1} = 2(k_1 + k_2 + \cdots + k_n) + 1$. Letting $n \to n - 1$ gives $k_n = 2(k_1 + k_2 + \cdots + k_{n-1}) + 1$. Subtracting this from the preceding relation gives $k_{n+1} - k_n = 2k_n \Longrightarrow k_{n+1} = 3k_n$. And since k_1 must be 1 (because that is the only way to verify a weight of 1 with one fixed weight), we inductively see that the k_i's are powers of 3. So the n fixed weights run from $3^0 = 1$ up to 3^{n-1}, and they can verify all weights up to their (geometric series) sum of $(3^n - 1)/2$, which is the upper bound we found in part (a). ♣

EXTENSION: You wish to pick n pairs of equal weights (for example, if $n = 3$, you might pick weights of 1, 1, 4, 4, 11, 11) such that you can verify any positive integer weight up to W. How should you choose the weights in order to maximize W? (Try to solve this before reading further.)

Along the lines of the reasoning in part (a), we observe that each pair of weights can be used in five ways: We can put (1) both weights on the left side, (2) one on the left and none on the right, (3) one on each side or use neither, (4) one on the right and none on the left, or (5) both on the right.

Therefore, the weights may be used in 5^n ways. As above, however, there are duplicates among these 5^n combinations. For every positive number, there is its negative, which represents the same net weight on the balance scale. Since it is, in principle, possible for the number 0 not to be repeated, an upper bound on the number of positive integer weights that can be verified with n pairs of fixed weights is $(5^n - 1)/2$.

It is indeed possible to achieve $W = (5^n - 1)/2$, by choosing the weights to be powers of 5. The weights should be 1, 1, 5, 5, 25, 25, etc. Powers of 5 have the relevant property that $2(1 + 5 + 5^2 + \cdots + 5^{n-1}) + 1 = 5^n - 2(1 + 5 + 5^2 + \cdots + 5^{n-1})$, as you can check. For example, if $n = 3$ we see that 63, which is not quite obtainable with three pairs of weights (which yield up to $2(1 + 5 + 25) = 62$), can be obtained by taking a weight from the fourth pair, namely 125, and subtracting off the highest possible sum obtainable with three pairs, namely 62. (And you can verify that with the fourth pair, all the intermediate weights up to $2 \cdot 125 + 62 = (5^4 - 1)/2$ are obtainable.) Alternatively, you can make a quick modification to the k_n argument we used in the preceding remark.

In general, if we use n k-tuples of weights, it is possible to verify any positive integer weight up to $[(2k + 1)^n - 1]/2$. The weights should be powers of $(2k + 1)$.

4. **Counterfeit coin**

(a) There are three possible outcomes of each weighing on the balance scale: left side heavier, right side heavier, or the two sides equal. In order to perform the given task in as few weighings as possible, we will need as much information as possible from each weighing. Hence, all three possible outcomes should be realizable for each weighing (except for the final weighing in some scenarios, as we will see below). So, for example, an initial weighing of six coins against six coins is probably not a good idea, because it isn't possible for the scale to balance. We should expect to have to switch coins from one side of the scale to the other, from one weighing to the next, in order to make the three possibilities realizable for a given weighing. Here is one scheme that does the task in three weighings:

Weigh four coins (labelled A_1, A_2, A_3, A_4) against four others (B_1, B_2, B_3, B_4). Let the remaining four be labelled C_1, C_2, C_3, C_4. There are three possible outcomes to this weighing:

1. The A group is heavier than the B group. We know in this case that the C coins are "good," and the "bad" coin is either an A or a B. If the bad coin is an A, it is heavy. If the bad coin is a B, it is light. For the second weighing, weigh (A_1, A_2, B_1) against (A_3, A_4, B_2). There are three possible outcomes:

    - If the (A_1, A_2, B_1) side is heavier, the bad coin must be A_1, A_2, or B_2. Weigh A_1 against A_2. If A_1 is heavier, it is the bad (heavy)

coin; if A_2 is heavier, it is the bad (heavy) coin; if they are equal, B_2 is the bad (light) coin.
- If the (A_3, A_4, B_2) side is heavier, the bad coin must be A_3, A_4, or B_1. Use the same strategy as in the previous case.
- If they are equal, the bad coin must be B_3 or B_4. Simply weigh them against each other; the light coin is the bad one.

2. The B group is heavier than the A group. This case is the same as the previous one, with A and B switched.

3. The A and B groups balance. So the bad coin is a C. For the second weighing, weigh (C_1, C_2) against $(C_3,$ good coin from A or $B)$. There are three possible outcomes:
- If the (C_1, C_2) side is heavier, weigh C_1 against C_2. If C_1 is heavier, it is the bad (heavy) coin; if C_2 is heavier, it is the bad (heavy) coin; if they are equal, C_3 is the bad (light) coin.
- If the (C_1, C_2) side is lighter, this is equivalent to the previous case, with "heavy" switched with "light."
- If they are equal, the bad coin is C_4. Weigh C_4 against a good coin to determine if it is heavy or light.

(b) **Lemma:** *Let there be N coins, about which our information is the following: The N coins may be divided into two sets, $\{H\}$ and $\{L\}$, such that i) if a coin is in $\{H\}$ and it turns out to be the bad coin, it is heavy; and ii) if a coin is in $\{L\}$ and it turns out to be the bad coin, it is light. Then given n weighings, the maximum value of N for which we can identify the bad coin, and also determine whether it is heavy or light, is $N = 3^n$.*

Proof: For the $n = 0$ case, the lemma is certainly true, because we have only $N = 3^0 = 1$ coin, and by assumption we know which of the two sets, $\{H\}$ and $\{L\}$, the one coin is in. So we can solve the problem for $N = 1$ coin with $n = 0$ weighings. And this is the maximum N, because we certainly can't solve the problem for $N = 2$ coins with $n = 0$ weighings.

We will now show by induction that the lemma is true for all n. That is, we'll assume that the lemma is true for n weighings and then show that it is also true for $n + 1$ weighings. We'll do this by first showing that $N = 3^{n+1}$ is solvable with $n + 1$ weighings, and then showing that $N = 3^{n+1} + 1$ is not always solvable with $n + 1$ weighings.

By assumption, the $N = 3^{n+1} = 3 \cdot 3^n$ coins are divided into $\{H\}$ and $\{L\}$ sets. On each side of the scale, put h coins from $\{H\}$ and l coins from $\{L\}$, with $h + l = 3^n$. (In general, there are many ways to do this. Either h or l may be zero, if necessary.) There are then 3^n coins left over.

There are three possible outcomes to this weighing:
- If the left side is heavier, the bad coin must be one of the h H-type coins from the left or one of the l L-type coins from the right.
- If the right side is heavier, the bad coin must be one of the h H-type coins from the right or one of the l L-type coins from the left.

- If the scale balances, the bad coin must be one of the 3^n leftover coins.

In each of these cases, the problem is reduced to a setup with 3^n coins that are divided into $\{H\}$ and $\{L\}$ sets. But this is assumed to be solvable with n weighings (by the inductive hypothesis), which means that the original set of 3^{n+1} coins is solvable with $n+1$ weighings. Therefore, since $N = 3^0 = 1$ is solvable for $n = 0$, and since we have just demonstrated that the induction step is valid, we conclude that $N = 3^n$ is solvable for all n.

Let us now show that $N = 3^{n+1} + 1$ is not always solvable with $n+1$ weighings. Assume inductively that $N = 3^n + 1$ is not always solvable with n weighings. ($N = 3^0 + 1 = 2$ is certainly not solvable for $n = 0$.) For the first weighing, the leftover pile can have at most 3^n coins in it, because the bad coin might end up being there. There must therefore be at least $2 \cdot 3^n + 1$ total coins on the scale, which then implies that there must be at least $2 \cdot 3^n + 2$ total coins on the scale, because the number must be even if we want to have a chance of gaining any information.[2] Depending on how the $\{H\}$ and $\{L\}$ coins are distributed on the scale, the first weighing will (assuming the scale doesn't balance) tell us that the bad coin is either in a subset containing s coins (say, the left H's and the right L's, if the left side is heavier) or in the complementary subset containing $2 \cdot 3^n + 2 - s$ coins (the right H's and the left L's, if the right side is heavier). One of these sets will necessarily have at least $3^n + 1$ coins in it, which by the inductive hypothesis is not necessarily solvable with n weighings. Of course, we might get lucky and end up with numbers that are solvable, but what we've shown here is that there is no guarantee of this. ∎

Returning to the original problem, let us first consider a modified setup where we have an additional known good coin at our disposal.

Claim: *Given N coins and W weighings, and given an additional known good coin, the maximum value of N for which we can identify the bad coin, and also determine whether it is heavy or light, is $N_W^{\text{g}} = (3^W - 1)/2$, where the superscript "g" signifies that we have a known good coin available.*

Proof: The claim is true for $W = 1$, because we simply need to weigh the $N_1^{\text{g}} = (3^1 - 1)/2 = 1$ coin against the known good coin; this will determine whether our coin is heavy or light. And $N_1^{\text{g}} = 1$ is indeed the maximum number of coins we can deal with in $W = 1$ weighings, because the 2-coin case isn't solvable, as you can quickly verify.

Assume inductively that the claim is true for W weighings. We will show that it is then true for $W + 1$ weighings. In the first of our $W + 1$ weighings, we can have (by the inductive hypothesis) at most $N_W^{\text{g}} = (3^W - 1)/2$ leftover coins not involved in the weighing, because the bad coin might end up

---

[2]To be picky, we could in fact gain information with an odd number of coins, in some very unlikely scenarios. For example, if we have k coins on one side and $k+1$ on the other, and if the scale balances, then we know that either the bad coin is on the $k+1$ side and it has a weight of zero, or it is on the k side and it has a weight twice that of a good coin. But we of course can't count on this unlikely scenario being realized.

being there (in which case we have many good coins at our disposal from the scale).

From the above lemma, we can have at most 3^W suspect coins on the scale. We can indeed have this many, if we bring in the additional known good coin to make the number of weighed coins, $3^W + 1$, be even (so that we can have an equal number on each side). If the scale doesn't balance, the 3^W suspect coins satisfy the hypotheses of the above lemma (they can be divided into $\{H\}$ and $\{L\}$ sets). So if the bad coin is among these 3^W coins, it can be determined in W additional weighings.

Therefore,

$$N^g_{W+1} = N^g_W + 3^W = \frac{3^W - 1}{2} + 3^W = \frac{3^{W+1} - 1}{2}, \qquad (3.1)$$

as we wanted to show. And we can't do any better than this value of N^g_{W+1}, because the N^g_W and 3^W sub-parts of it are the maximal numbers for the respective sub-cases (N^g_W by the inductive hypothesis, and 3^W by the above lemma). ∎

We can now finally solve our original problem, with this corollary:

Corollary: *Given N coins and W weighings (and not having an additional known good coin available), the maximum value of N for which we can identify the bad coin, and also determine whether it is heavy or light, is*

$$N^{ng}_W = \frac{3^W - 1}{2} - 1, \qquad (3.2)$$

where the superscript "ng" signifies that we do not have a known good coin available.

Proof: If we are not given a known good coin, the only modification to the reasoning in the above claim is that we can't put 3^W suspect coins on the scale (if we want to gain any information), because 3^W is odd. (This is the only time we needed to use the *additional* known good coin in the above claim.) So we are limited to the (even) total of $3^W - 1$ coins on the scale, and we now obtain

$$N^{ng}_{W+1} = N^g_W + (3^W - 1) = \frac{3^W - 1}{2} + (3^W - 1) = \frac{3^{W+1} - 1}{2} - 1, \qquad (3.3)$$

as we wanted to show. And we can't do any better than this value of N^{ng}_{W+1}, because the N^g_W and $3^W - 1$ sub-parts of it are the maximal numbers for the respective sub-cases.

Remember that if the scale balances, so that we know the bad coin is a leftover coin, then from that point on, we do indeed have a known good coin at our disposal (any coin on the scale). So $N^g_W = (3^W - 1)/2$, as opposed to N^{ng}_W, is indeed what appears after the first "=" sign in Eq. (3.3). ∎

Comparing Eqs. (3.1) and (3.3), we see that N_W is decreased by 1 if we don't have a known good coin at the start. So $N_W^{ng} = (3^W - 1)/2 - 1$ is the final answer to part (b) of this problem. If $W = 3$, we obtain $N_3^{ng} = (3^3 - 1)/2 - 1 = 12$, consistent with the result in part (a).

REMARK: It is possible to determine an upper bound for N_W^g and N_W^{ng}, without going through all of the above work. (However, the reasoning here doesn't say anything about whether the upper bound is obtainable.) We can do this by considering the number of possible outcomes of the collective W weighings. There are three possibilities for each weighing (left side heavier, right side heavier, or the two sides equal), so there are at most 3^W possible outcomes. Each of these outcomes may be labelled by a string of W letters. For example, if $W = 5$ then one possible outcome/string is LLRER (with L for left, R for right, and E for equal).

However, the EEEEE string (where the scale always balances) doesn't give enough information to determine whether the bad coin is heavy or light. So we have at most $3^W - 1$ useful outcomes. Therefore, since there are two possibilities for each coin (heavy or light) in the event that it is the bad coin, we can have at most $(3^W - 1)/2$ coins. If we had more than $(3^W - 1)/2$, then at least one particular string would need to be associated with at least two different conclusions. For example, LLRER might be associated with both "coin #8 is the bad coin and it is light" and "coin #13 is the bad coin and it is heavy." But then we haven't solved the problem (by reaching a unique conclusion).

Note that although there may be different possibilities for coin placement at various points in the weighing process (depending on who is deciding what coins to use and where they go on the scale), only 3^W possible outcomes are realizable for a given overall strategy. A given strategy involves writing down an "if, then" tree that specifies which coins you will put on each side of the scale at each point in a tree like the one shown in Fig. 3.4. Once you pick a strategy, there are at most 3^W possible outcomes, and hence at most $3^W - 1$ useful outcomes. For example, with four weighings, the R at the bottom of Fig. 3.4 is the end of the ERLR string. If the four weighings result in this string, then we might have enough information to conclude that, say, coin #8 is the bad coin and that it is light.

As we saw above, the upper bound of $(3^W - 1)/2$ is obtainable if we have an additional known good coin at our disposal. But it turns out that we fall short of the bound by 1 if we do not have an additional known good coin. ♣

5. The game of Nim

As with many problems, this one can be solved in two possible ways. We can (a) write down the correct answer, through some stroke of genius, and then verify that it works, or (b) work out some simple cases, get a feel for the problem, and eventually wind our way around to the correct answer. For the problem at hand, let's proceed via the second method and try to arrive at the answer with some motivation.

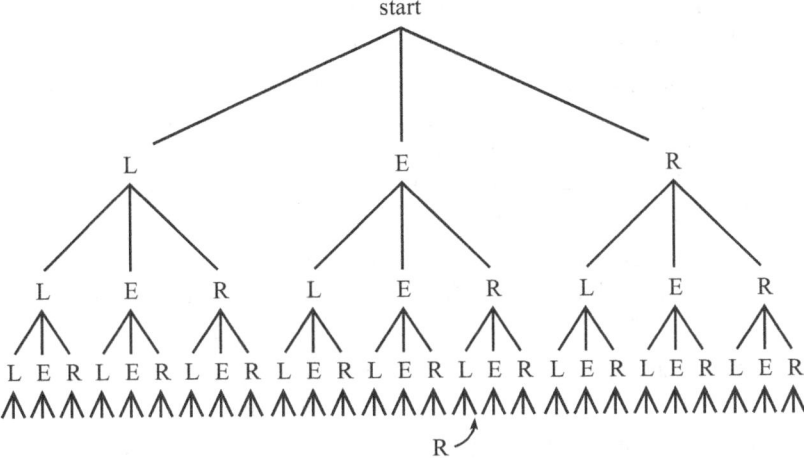

Figure 3.4

We'll start by working out what happens in particular cases of small numbers of coins in the piles, and then we'll look for a pattern. A reasonable way to organize the results is to determine which combinations of numbers are guaranteed losing positions, assuming that both players are aware of the optimal strategy. (We could of course look at winning positions instead. But losing positions are more convenient, for reasons we'll see.)

The most obvious guaranteed losing position (LP) is piles with coins of numbers $(1, 1, 0)$. If you encounter this setup, you must pick one coin, and then your opponent will pick the last one and thereby win. More generally, the triplet $(N, N, 0)$ is an LP, because if you take n coins from one pile, your opponent will take n coins from the other. She will keep matching you on each turn, until finally the triplet is $(0, 0, 0)$, with the last coin(s) having been removed by her. Note that triplets of the form $(N, M, 0)$ and (N, N, M) are therefore winning positions (WP), because it is possible to turn them into an LP with one move. (Remove $|N - M|$ coins from the larger pile in the first case, and all M coins from the last pile in the second case.)

The above reasoning utilizes the following two properties of an LP: (1) Removal of any number of coins from one pile of an LP creates a non-LP (a WP), and (2) since we then have a WP, it is always possible on the following turn to bring the triplet back to an LP.

Consider now the cases where no two piles have the same numbers of coins. We'll start with $(1, 2, x)$. Given the $(N, M, 0)$ and (N, N, M) WP's we noted above, $(1, 2, 3)$ is the first possibility for an LP. And we quickly see that it is indeed an LP, because the removal of any number of coins from any one of the piles yields a triplet of the form $(N, M, 0)$ or (N, N, M), which are WP's.

Note that once we have found an LP, we know that any triplet that has two numbers in common with the LP, with its third number larger than the corresponding one in the LP, must be a WP. This is true because it is possible to turn the triplet into an LP by removing coins from the third pile.

If we look at other triplets in which the first number is 1, we find that (1, 4, 5), (1, 6, 7), (1, 8, 9), etc., are LP's, as you can check by showing that any move turns them into a WP. If you're wondering whether we missed any LP's of the form (1, 3, x) with $x > 2$, the answer is no, because the removal of coins from the x pile can turn the triplet into the LP triplet (1, 3, 2) we found above.

Now consider cases where 2 is the smallest number of coins in a pile. We find, after a little fiddling, that (2, 4, 6), (2, 5, 7), (2, 8, 10), and (2, 9, 11), etc., are LP's, as you can check.

Similar fiddling, starting with 3, gives (3, 4, 7), (3, 5, 6), (3, 8, 11), and (3, 9, 10), etc., as LP's.

Let's now make a table of this hodgepodge of results, for up to seven coins in a pile; see Table 3.2. The two axes are the first two numbers in an LP triplet, and the entry in the table is the third. The table is of course symmetric.

	0	1	2	3	4	5	6	7
0	0	1	2	3	4	5	6	7
1	1	0	3	2	5	4	7	6
2	2	3	0	1	6	7	4	5
3	3	2	1	0	7	6	5	4
4	4	5	6	7	0	1	2	3
5	5	4	7	6	1	0	3	2
6	6	7	4	5	2	3	0	1
7	7	6	5	4	3	2	1	0

Table 3.2: LP triplets, for up to seven coins.

As a first guess at the key to this table, we might say that two numbers in an LP triplet must add up to the third. This, however, does not work for the (3, 5, 6) triplet. It also does not work for the (3, 9, 10) triplet we found above. Continuing on to higher numbers, the guess seems to work for triplets starting with 4. But then if we start with 5, we eventually find the LP triplets (5, 9, 12) and (5, 11, 14), for which the sum of two numbers doesn't equal the third.

In an effort to find the key, let us exploit the patterns in Table 3.2, perhaps brought out best by the grouping in Table 3.3. The entries in the upper right 4×4 box are 4 more than the corresponding entries in the upper left box. Likewise, within each 4×4 box, the entries in the upper right 2×2 box are 2 more than the entries in the upper left box. Similar results would be evident if we doubled the size of the box (out to 15), where we would see 8×8 boxes having entries differing by 8.

All of this suggests that powers of 2 are important in this problem. We therefore should consider writing the numbers in a way where factors of 2 are evident, that is, in base 2. There is no guarantee that this will help, but let's try it and see what

0	1	2	3	4	5	6	7
1	0	3	2	5	4	7	6
2	3	0	1	6	7	4	5
3	2	1	0	7	6	5	4
4	5	6	7	0	1	2	3
5	4	7	6	1	0	3	2
6	7	4	5	2	3	0	1
7	6	5	4	3	2	1	0

Table 3.3: LP triplets, grouped in a helpful manner.

happens. Table 3.4 shows the troublesome LP triplets we've found (the ones for which two of the numbers don't add up to the third), written in base 2.

3:	11	3:	11	5:	101	5:	101
5:	101	9:	1001	9:	1001	11:	1011
6:	110	10:	1010	12:	1100	14:	1110

Table 3.4: A few LP's for which two of the numbers don't add up to the third.

What property do these triplets have? When written in the above form, we see that each column in base 2 contains an even number of 1's. After checking some other triplets, this appears to be true in general for an LP. We will prove this with the following theorem.

Theorem: *Call a triplet an E-triplet (the "E" stands for "even") if it has the following property: When the three numbers are written in base 2, there is an even number of (that is, either zero or two) 1's in each digit's place (each column in the lists in Table 3.4). Then a triplet is a losing position (LP) if and only if it is an E-triplet.*

Proof: Let us establish the following three facts concerning E-triplets:

(a) Removal of any number of coins from any single pile of an E-triplet turns the triplet into a non-E-triplet.

(b) Given a non-E-triplet, it is always possible to remove coins from a single pile to turn the triplet into an E-triplet.

(c) (0,0,0) is an E-triplet.

These facts may be demonstrated as follows:

(a) This fact is true because any two numbers in an E-triplet uniquely determine the third. (Two 1's in a column mean that the third number must be a 0. One 0 and one 1 in a column mean that the third number must be a 1. And two 0's in a column mean that the third number must be a 0. A blank space counts as a 0.) So changing any one of the numbers changes it from the unique number determined by the other two.

(b) We can turn any non-E-triplet into an E-triplet in the following way. Write the three numbers of coins in base 2, and put them on top of each other, with the unit's digits aligned, as we did in Table 3.4. Starting from the *left*, look at each digit's column until you find a column with an odd number of (that is, either one or three) 1's. Let this be the nth column (counting from the *right*).

If there is one 1 in the nth column, label the number containing this 1 as A. If there are three 1's, arbitrarily pick any of the three numbers to be A. Remove coins from A by switching the 1 in the nth column to a 0, and also by switching any 1's to 0's, or 0's to 1's, in other columns to the right of the nth column, in order to produce an even number of 1's in all columns. We have now created an E-triplet.

Note that this switching of 1's and 0's does indeed correspond to removing (as opposed to adding) coins from A, because even if all of the columns to the right of the nth column involve switching 0's to 1's, this addition of $1 + 2 + 4 + \cdots + 2^{n-2} = 2^{n-1} - 1$ coins is still less than the subtraction of the 2^{n-1} coins arising from the 1-to-0 switch in the nth column. This is why we went through the above procedure of identifying A.

As a concrete example of this process, consider the triplet $(5, 14, 22)$. This is a non-E-triplet, because some (actually most, in this case) of the columns have an odd number of 1's:

```
 5:      101
14:     1110
22:    10110
```

Following the above procedure, the fifth column (counting from the right) is the first one (starting from the left) that has an odd number of 1's. There is only one 1 in this column, so 22 is our A. We therefore change the 1 in the fifth column to a zero. And then we make the other changes to $A = 22$ shown below in bold, to yield an even number of 1's in each column. The second column is the only one that doesn't involve a change. We end up with the E-triplet $(5, 14, 11)$, which is the last of the triplets listed in Table 3.4.

```
 5:      101
14:     1110
11:    01011
```

(c) This third fact is true, by the definition of an E-triplet.

The first two of the above facts show that if player X receives an E-triplet on a given turn, then player Y can ensure that X receives an E-triplet on every subsequent turn. Therefore, X must always *create* a *non*-E-triplet, by the first of the three facts. X therefore cannot take the last coin(s) (and thereby win), because he cannot create the E-triplet $(0, 0, 0)$. Therefore, an E-triplet is a losing position. And conversely, a non-E-triplet is a winning position, because it can always be turned into a (losing) E-triplet, due to the second of the above facts. ■

The best strategy in this game is therefore to give your opponent an E-triplet whenever you can. If both players are aware of this strategy, then the outcome is determined by the initial piles of coins. If they form an E-triplet, then the player who goes first loses. If they do not form an E-triplet, then the player who goes first wins, because he can always create an E-triplet to give to his opponent.

REMARKS: If the starting numbers of coins are random, then the player who goes first will most likely win, because most triplets are non-E-triplets. We can demonstrate this fact by making the somewhat crude assumption that the three numbers are random numbers from 0 to $2^n - 1$, that is, they each have n digits in base 2 (many of which may be zero). There are then $(2^n)^3$ possible triplets. But there are only 4^n possible E-triplets, because each of the n columns of three digits (when we write the three E-triplets on top of each other) must take one of the following four forms:

$$\begin{pmatrix} 0 \\ 0 \\ 0 \end{pmatrix}, \quad \begin{pmatrix} 1 \\ 1 \\ 0 \end{pmatrix}, \quad \begin{pmatrix} 1 \\ 0 \\ 1 \end{pmatrix}, \quad \begin{pmatrix} 0 \\ 1 \\ 1 \end{pmatrix}. \tag{3.4}$$

The fraction of E-triplets is therefore $4^n/2^{3n} = 1/2^n$, which goes to zero for large n. (Equivalently, 4 out of the $2^3 = 8$ possible forms for each column are E-triplet forms. So the probability that all n columns are of E-triplet form is $(4/8)^n = 1/2^n$.)

Note that there is nothing special about having three piles. We can have any number of piles (but still two players), and all of the above reasoning still holds. Losing positions are ones that have an even number of 1's in each column when written in base 2. The three facts in the above theorem still hold. ♣

6. **Monochromatic triangle**

   (a) Our strategy will be to try to avoid forming a monochromatic triangle, and to then show that this task is impossible. Consider one of the points and the 16 lines drawn from it to the other 16 points. Since we have three colors, the pigeonhole principle (or rather a standard generalization of it)[3] implies that at least six of these lines must be of the same color. (Otherwise we would be able to color at most $3 \cdot 5 = 15$ lines.) Let this color be red.

   Now consider the six points at the ends of these red lines. Look at the lines going from one of these points to the other five. In order to not form a red triangle, each of these five lines must be either green or blue. Hence (by the pigeonhole principle) at least three of them must be of the same color. (Otherwise we would be able to color at most $2 \cdot 2 = 4$ lines.) Let this color be green.

   Finally, consider the three points at the ends of the three green lines. If any one of the three lines connecting them is red, a red triangle is formed. And if any one of the three lines connecting them is green, a green triangle is

---

[3]The most basic form of the pigeonhole principle says that if you have n pigeons and $n-1$ pigeonholes, then at least two pigeons must go in one pigeonhole.

formed. Therefore, they must all be blue, which means that a blue triangle is formed. Hence there is no way to avoid forming a monochromatic triangle.

(b) Consider the problem in the case of $n = 4$, in order to get an idea of how the solution generalizes. We claim that 66 points necessitate a monochromatic triangle. As in the case of $n = 3$, consider one of the points and the 65 lines drawn from it to the other 65 points. Since we have four colors, the pigeonhole principle implies that at least 17 of these lines must be of the same color. (Otherwise we would be able to color at most $4 \cdot 16 = 64$ lines.) In order to not form a monochromatic triangle, the lines joining the endpoints of these 17 lines must use only the remaining three colors. So the problem reduces to the $n = 3$ case with 17 points we dealt with in part (a), where we showed that a monochromatic triangle is necessarily formed. Generalizing this reasoning yields the following result:

Claim: *If n colors and P_n points necessitate a monochromatic triangle, then $n + 1$ colors and*

$$P_{n+1} = (n + 1)(P_n - 1) + 2 \tag{3.5}$$

points also necessitate a monochromatic triangle.

Proof: Given $n + 1$ colors and $(n + 1)(P_n - 1) + 2$ points, consider one of the points and the $(n + 1)(P_n - 1) + 1$ lines drawn from it to the other points. Since we have $n + 1$ colors, the pigeonhole principle implies that at least P_n of these lines must be of the same color. (Otherwise we would be able to color at most $(n + 1)(P_n - 1)$ lines.) In order to not form a monochromatic triangle, the lines joining the endpoints of these P_n lines must use only the remaining n colors. But by the hypothesis of the claim, there must then be a monochromatic triangle. ∎

We can now use the recursion relation in Eq. (3.5) to find P_n as a function on n. Our starting value will be $P_1 = 3$; three points connected by three lines of a single color will certainly form a monochromatic triangle. Of course, so will four or five, etc., points connected with a single color. But our goal here is to be as thrifty as we can with the number of points.

First, let's replace $n + 1$ with n for convenience, which turns Eq. (3.5) into $P_n = n(P_{n-1} - 1) + 2$. If we repeatedly plug this recursion relation into itself (that is, if we write P_{n-1} in terms of P_{n-2}, and then write P_{n-2} in terms of P_{n-3}, and so on), a pattern will emerge. However, things are a little less messy if we make one more modification and rewrite the recursion relation as $P_n - 1 = n(P_{n-1} - 1) + 1$, which can be expressed as

$$Q_n = nQ_{n-1} + 1, \tag{3.6}$$

where $Q_n \equiv P_n - 1$. Our starting value of $P_1 = 3$ implies $Q_1 = 2$. Alternatively, we can formally start the sequence with $Q_0 = 1$ (which correctly yields $Q_1 = 2$ in Eq. (3.6)). Let's now repeatedly plug the

$Q_n = nQ_{n-1} + 1$ relation into itself. After three iterations we obtain

$$Q_n = nQ_{n-1} + 1 \tag{3.7}$$
$$= n(n-1)Q_{n-2} + n + 1$$
$$= n(n-1)(n-2)Q_{n-3} + n(n-1) + n + 1$$
$$= n(n-1)(n-2)(n-3)Q_{n-4} + n(n-1)(n-2) + n(n-1) + n + 1$$
$$= n!\left(\frac{Q_{n-4}}{(n-4)!} + \frac{1}{(n-3)!} + \frac{1}{(n-2)!} + \frac{1}{(n-1)!} + \frac{1}{n!}\right). \tag{3.8}$$

If we keep iterating until the "$n-4$" here becomes a zero, we obtain

$$Q_n = n!\left(\frac{Q_0}{0!} + \frac{1}{1!} + \frac{1}{2!} + \cdots + \frac{1}{(n-1)!} + \frac{1}{n!}\right). \tag{3.9}$$

Using the fact that $Q_0 = 1$ and recalling the definition $Q_n \equiv P_n - 1$, we arrive at the value of P_n:

$$P_n = n!\left(1 + \frac{1}{1!} + \frac{1}{2!} + \cdots + \frac{1}{(n-1)!} + \frac{1}{n!}\right) + 1. \tag{3.10}$$

You can quickly double check that the above Q_n and P_n expressions satisfy their respective recursion relations.

The sum in the parentheses in Eq. (3.10) is smaller than e (which equals $\sum_0^\infty 1/k!$) by a margin that is less than $1/n!$. So $n!$ times the sum is smaller than $n!e$ by a margin that is less than 1. Therefore, P_n (which includes the "+1" in Eq. (3.10)) is equal to the smallest integer greater than $n!e$, as we wanted to show. (Note that although $\lceil a \rceil$ is defined to be the smallest integer greater than or equal to a, the "or equal to" possibility isn't relevant here, because $n!e$ can never be an integer, since e is irrational.)

REMARK: For $n = 1, 2, 3$, we know from the above claim that the numbers $\lceil n!e \rceil$ (which equal 3, 6, 17, respectively) necessitate a monochromatic triangle. It turns out that additionally these are the *smallest* numbers of points that necessitate a monochromatic triangle. This is true because for $n = 1$, two points don't even form a triangle. And for $n = 2$, you can easily construct a pentagon that doesn't contain a monochromatic triangle. For $n = 3$, things are much more difficult, but in 1955 Greenwood and Gleason showed that 16 points do *not* necessitate a monochromatic triangle. (See R. E. Greenwood and A. M. Gleason, "Combinatorial Relations and Chromatic Graphs," Canadian J. Math, 7 (1955), 1–7.) For $n \geq 4$, the problem of finding the smallest number of points that necessitate a monochromatic triangle is unsolved, I believe. But we can at least say that $\lceil n!e \rceil$ is an upper bound on the smallest number. ♣

7. **AM-GM Inequality**

The three suggested steps are:

- Show that I_2 is true: We'll use the fact that the square of any real number is greater than or equal to zero. In particular, if the x_i are non-negative real numbers, then

$$\left(\sqrt{x_1} - \sqrt{x_2}\right)^2 \geq 0$$
$$\implies x_1 - 2\sqrt{x_1 x_2} + x_2 \geq 0$$
$$\implies \frac{x_1 + x_2}{2} \geq \sqrt{x_1 x_2}, \tag{3.11}$$

as desired. And from the first line above, we see that equality holds if and only if $x_1 = x_2$.

- Show that I_n implies I_{2n}: We'll do this by dividing a set of $2n$ numbers into two sets of n numbers, and then invoking both I_2 and I_n. For convenience, define $S_{a,b}$ as the sum of x_a through x_b, and define $P_{a,b}$ as the product of x_a through x_b. Then under the assumption that I_n is true, we have

$$\frac{S_{1,n}}{n} \geq \sqrt[n]{P_{1,n}} \quad \text{and} \quad \frac{S_{n+1,2n}}{n} \geq \sqrt[n]{P_{n+1,2n}}. \tag{3.12}$$

We'll now apply I_2 to the two quantities $S_{1,n}/n$ and $S_{n+1,2n}/n$. This gives (using Eq. (3.12) to obtain the second line)

$$\frac{1}{2}\left(\frac{S_{1,n}}{n} + \frac{S_{n+1,2n}}{n}\right) \geq \sqrt{\frac{S_{1,n}}{n} \cdot \frac{S_{n+1,2n}}{n}}$$
$$\geq \sqrt{\sqrt[n]{P_{1,n}} \cdot \sqrt[n]{P_{n+1,2n}}}$$
$$\implies \frac{S_{1,2n}}{2n} \geq \sqrt[2n]{P_{1,2n}}, \tag{3.13}$$

which is the I_{2n} statement. So I_n implies I_{2n}. As an exercise, you can show inductively that equality holds if and only if all $2n$ numbers are equal.

- Show that I_n implies I_{n-1}: If we are assuming that I_n holds, then it holds for *any* set of n non-negative numbers. In particular, it holds if x_n equals the geometric mean of the other $n-1$ numbers, that is, if $x_n = \sqrt[n-1]{P_{1,n-1}}$. A valid I_n statement is then

$$\frac{S_{1,n-1} + \sqrt[n-1]{P_{1,n-1}}}{n} \geq \sqrt[n]{P_{1,n-1} \cdot \sqrt[n-1]{P_{1,n-1}}}$$
$$\implies S_{1,n-1} + \sqrt[n-1]{P_{1,n-1}} \geq n \cdot \sqrt[n-1]{P_{1,n-1}}$$
$$\implies \frac{S_{1,n-1}}{n-1} \geq \sqrt[n-1]{P_{1,n-1}}, \tag{3.14}$$

which is the I_{n-1} statement. Again, equality holds if and only if all $n-1$ numbers are equal. As an exercise, you can also show that I_n implies I_{n-1} by letting x_n equal the arithmetic mean of the other $n-1$ numbers, instead of the geometric mean.

Putting together the above results, we see that because I_2 is true, the "I_n implies I_{2n}" statement implies that $I_4, I_8, I_{16}, \ldots$ are all true. And then the "I_n implies I_{n-1}" statement implies that we can work backwards from any power of 2 to show that I_n holds for all n less than that power. And since any integer has a power of 2 larger than it, we see that I_n is true for all n.

REMARKS:

1. The AM-GM inequality provides a way to solve certain maximization problems without using calculus. For example, let's say we want to maximize xy^2, subject to the constraint $x + y = 1$, assuming that both x and y are positive.

   The calculus method is to plug $x = 1 - y$ into xy^2, yielding $(1 - y)y^2$. Setting the derivative equal to zero gives $2y - 3y^2 = 0 \implies y = 2/3$ (or $y = 0$), and so $x = 1/3$. The maximum value of xy^2 (for positive x and y) is then $4/27$.

   The AM-GM-inequality method is to say that

   $$1 = x + y = x + \frac{y}{2} + \frac{y}{2} \geq 3 \cdot \sqrt[3]{x \cdot \frac{y}{2} \cdot \frac{y}{2}}$$
   $$\implies \frac{1}{3} \geq \sqrt[3]{\frac{xy^2}{4}} \implies \frac{4}{27} \geq xy^2, \qquad (3.15)$$

   in agreement with the calculus method. Equality holds when the three numbers in the AM-GM inequality are equal, that is, when $x = y/2$. Since $x + y = 1$, this implies $x = 1/3$ and $y = 2/3$, as above.

   The motivation for using the $y/2$ quantities in Eq. (3.15) is that the AM-GM inequality tells us that $(ax) + (by) + (cy) \geq 3 \cdot \sqrt[3]{(ax)(by)(cy)}$, for any non-negative values of the various parameters. The righthand side involves the product xy^2. So we just need to pick a, b, and c so that the lefthand side looks like $x + y$. Hence $a = 1$ and $b = c = 1/2$. Technically, any b and c values satisfying $b + c = 1$ will work, but equality in the AM-GM inequality is achieved only if all the numbers involved are equal. That's why we picked b and c to be equal, and hence equal to $1/2$. Picking other values for b and c (with $b + c = 1$) would produce a perfectly valid inequality for xy^2. It's just that equality would never occur, so we wouldn't be able to say what the maximum value of xy^2 is.

2. We can also prove the AM-GM inequality by using calculus. Consider n numbers, and let S and P be their sum and product (we won't bother labeling these as $S_{1,n}$ and $P_{1,n}$, as we did above). We'll use induction, so we'll assume that the AM-GM inequality holds for n (that is, the inequality $S/n \geq P^{1/n}$ is true), and then we'll show that the inequality also holds for $n + 1$. (You should try to work this out before reading further.)

   Let the $(n+1)$th number be x. Our goal is to show that

   $$\frac{S + x}{n + 1} - (Px)^{1/(n+1)} \geq 0. \qquad (3.16)$$

We'll do this by calculating the minimum value of the lefthand side (as a function of x) and showing that it is greater than or equal to zero. Setting the derivative with respect to x equal to zero gives

$$\frac{1}{n+1} - \frac{P^{1/(n+1)} x^{-n/(n+1)}}{n+1} = 0 \implies x = P^{1/n}. \qquad (3.17)$$

(You can quickly see that the second derivative is positive, so we do indeed have a minimum.) Plugging this value of x back into the lefthand side of Eq. (3.16) yields a minimum value of

$$\frac{S + P^{1/n}}{n+1} - \left(P \cdot P^{1/n}\right)^{1/(n+1)} = \frac{S}{n+1} + P^{1/n}\left(\frac{1}{n+1} - 1\right)$$

$$= \frac{n}{n+1}\left(\frac{S}{n} - P^{1/n}\right). \qquad (3.18)$$

But this quantity is greater than or equal to zero by our inductive hypothesis, $S/n \geq P^{1/n}$. Eq. (3.16) therefore holds, and the inductive step is complete. And since the AM-GM inequality is trivially true for $n = 1$ (where equality holds), we see that it is true for all n. Furthermore, equality holds if and only if $x = P^{1/n}$ at every inductive step, otherwise the minimum possible value (namely zero) we found in Eq. (3.18) wouldn't be obtained. But if $x = P^{1/n}$ at every inductive step, then all n numbers must be equal. This is therefore the condition for equality in the AM-GM inequality. ♣

8. **Crawling ant**

At time t, the movable end of the rubber band (let's label this end as E) is a distance $\ell(t) = L + Vt$ from the wall. Let $F(t)$ be the ratio of the distances:

$$F(t) = \frac{\text{distance from } E \text{ to ant}}{\text{distance from } E \text{ to wall}}. \qquad (3.19)$$

So $F(t)$ starts at zero, and it equals 1 when (or if) the ant reaches the wall. Our task is therefore to determine if $F(t)$ eventually equals 1, and if so, at what time t.

During a little time interval dt, the ant moves a distance $u\,dt$ with respect to the rubber band. The band has length $L + Vt$ at time t, so the fraction $F(t)$ increases by $u\,dt/(L + Vt)$. That is, $dF(t) = u\,dt/(L + Vt)$.

You might be worried that the length of the band changes from $L+Vt$ to $L+V(t+dt)$ during the interval dt. So there is an ambiguity in what length to use in the denominator of $dF(t)$. However, this ambiguity doesn't matter, because it would yield corrections only at second order in dt, since the numerator of $u\,dt/(L+Vt)$ is already first order in dt.

Integrating our expression for dF from $F = 0$ to $F = 1$ (to see if there is actually a t value that makes $F = 1$; call this t value t_w, with "w" for wall) gives

$$\int_0^1 dF = \int_0^{t_w} \frac{u\,dt}{L+Vt} \implies 1 = \frac{u}{V} \ln(L+Vt)\Big|_0^{t_w} \implies 1 = \frac{u}{V}\ln\left(1 + \frac{Vt_w}{L}\right). \qquad (3.20)$$

Solving for t_w gives

$$t_w = \frac{L}{V}\left(e^{V/u} - 1\right). \tag{3.21}$$

For large V/u, the time it takes the ant to reach the wall becomes exponentially large, but it does indeed reach it in a finite time, for any (nonzero) value of u. For small V/u, the Taylor approximation $e^x \approx 1 + x$ (see the appendix for a review of Taylor series) quickly reduces Eq. (3.21) to $t_w \approx L/u$, as it should; the ant essentially walks a distance L at speed u.

REMARKS:

1. This setup involving an ant crawling on a rubber band is a very helpful model for understanding a certain topic in physics/cosmology, namely the specifics of how light/photons travel in an expanding universe.[4] The rubber band itself represents space, the wall represents the earth, the other end E of the rubber band represents a distant galaxy, and the ant represents a photon emitted from the galaxy. The ant's speed u is then the speed of light, $c = 3 \cdot 10^8$ m/s, because that is the speed of a photon with respect to the local space it is traveling through. In the actual expanding universe we live in, the speed V of a galaxy isn't constant, but that can be taken into account without too much difficulty.

   The result of this problem tells us that even if $V > c$, that is, even if a galaxy is receding from the earth faster than the speed of light (yes, this is possible in an expanding universe), the photon will still eventually reach the earth, given enough time (as long as V doesn't increase too rapidly with time).

2. If $u < V$, the ant will initially get carried away from the wall before it eventually comes back and reaches the wall. What is the maximum distance the ant gets from the wall? (Try to solve this before reading further.)

   To find the maximum distance, we'll first find the functional form of $F(t)$. We can do this by simply letting the upper limit of the dF integral in Eq. (3.20) be $F(t)$ instead of 1. This gives

   $$F(t) = \frac{u}{V} \ln\left(1 + \frac{Vt}{L}\right). \tag{3.22}$$

   $F(t)$ represents the fractional distance from the moving end E to the ant. So the factional distance from the wall to the ant is $1 - F(t)$. The ant's distance from the wall is therefore

   $$x(t) = \left(1 - F(t)\right)(L + Vt) = \left(1 - \frac{u}{V}\ln\left(1 + \frac{Vt}{L}\right)\right)(L + Vt). \tag{3.23}$$

   Setting the derivative of $x(t)$ equal to zero gives, as you can verify,

   $$\left(1 - \frac{u}{V}\ln\left(1 + \frac{Vt}{L}\right)\right)V - u = 0. \tag{3.24}$$

---

[4] A nice introduction to the physics of an expanding universe is C. H. Lineweaver and T. M. Davis, "Misconceptions About the Big Bang," Scientific American, March 2005, 36–45.

Note that we could have arrived at this result by simply recognizing that the speed of the ant is zero at its maximum position. This means that the $(1 - F(t))V$ speed (of a dot painted on the rubber band near the ant) away from the wall due to the stretching (since the band stretches uniformly) cancels the u speed (of the ant relative to the dot on the band) toward the wall due to the crawling.

Solving Eq. (3.24) for t gives

$$t_{max} = \frac{L}{V}\left(e^{V/u-1} - 1\right) \qquad (V \geq u). \qquad (3.25)$$

This holds only if $V \geq u$, because it gives a negative t_{max} if $V < u$. If $V < u$, then $t_{max} = 0$ and $x_{max} = L$. That is, the maximum distance is simply the distance L right at the start, and the ant gets closer to the wall as time goes on. Plugging the t_{max} from Eq. (3.25) into Eq. (3.23) gives

$$x_{max} = \frac{u}{V}\frac{L}{e}e^{V/u} \qquad (V \geq u). \qquad (3.26)$$

If $u = V$, this correctly gives $x_{max} = L$. And if, for example, $V = 2u$, then $x_{max} \approx (1.36)L$. If $V = 10u$, then $x_{max} \approx (810)L$. Interestingly, for large V/u, the t_{max} in Eq. (3.25) is approximately $1/e$ times the time it takes the ant to reach the wall, given in Eq. (3.21). ♣

9. **Apple core**

If the core is very thin (compared with h), then the eaten part of the apple is approximately a whole sphere with a radius essentially equal to $h/2$. If in the other extreme the core is very wide (compared with h), then the (nearly) hemispherical bubbles on its top and bottom make up most of the apple, and all that was eaten is a long thin band going around the "equator" of the apple. It just so happens that in all cases, the volume of the eaten part exactly equals the volume of a sphere of radius $h/2$. So even though you might think there is missing information (namely, the radius R of the apple) in the statement of the problem, the answer is in fact independent of R. This can be shown as follows.

Let $\ell = h/2$, for convenience. Consider a cross section of the apple (produced by a horizontal plane) at a distance d (with $d < \ell$) above the center of the apple. This cross section is represented by the horizontal line GC in Fig. 3.5.

The eaten part of this cross section is an annulus with inner radius BD and outer radius BC. Right triangle ABC tells us that $BC = \sqrt{R^2 - d^2}$. And $BD = EF$, which from right triangle AEF has the value $BD = EF = \sqrt{R^2 - \ell^2}$. The area of the annulus is therefore $\pi(BC)^2 - \pi(BD)^2 = \pi(\ell^2 - d^2)$. This area, however, is exactly the same as the cross-sectional area of a sphere of radius ℓ, at a distance d above the center. The radius of the circular cross section is $\sqrt{\ell^2 - d^2}$; see Fig. 3.6.

Now, if all the corresponding cross-sectional areas of two objects are equal, then the two objects have the same volume. This is true because we could imagine

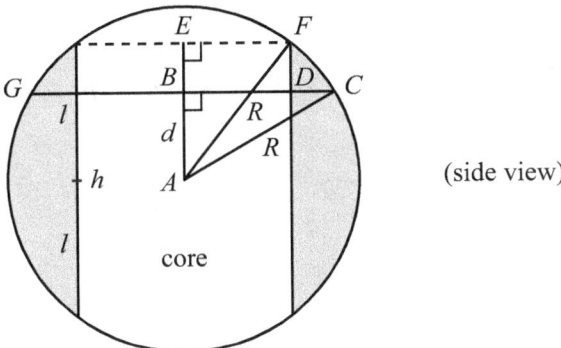

Figure 3.5

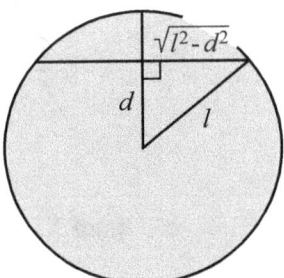

Figure 3.6

slicing the two objects into thin pancakes, all with the same tiny thickness (or at least with corresponding pancakes having the same thickness). Any two corresponding pancakes in the two objects have the same volume since they have the same cross-sectional area and thickness. So the two objects must have the same total volume. Therefore, the volume of the eaten part of the apple in Fig. 3.5 is the same as the volume of the sphere of radius $\ell = h/2$ in Fig. 3.6, which is $(4/3)\pi(h/2)^3 = \pi h^3/6$.

REMARK: In the limit where the core is very thin, the eaten part is (as we noted at the beginning of this solution) essentially the whole apple, which means that the volume of the eaten part is $(4/3)\pi(h/2)^3$. In the limit where the core is very wide, the eaten part is a long thin band going around the equator of the apple; the core is nearly the whole apple. In this limit, we can use a Taylor series to calculate (approximately) the volume of the eaten part and verify that it equals $(4/3)\pi(h/2)^3$. This approximate calculation will take longer than the exact calculation above, but it's still good to do. Try to work it out yourself before reading further.

In Fig. 3.7, the half-width of the core, AB, has length $\sqrt{R^2 - \ell^2}$. Using the Taylor series $\sqrt{1 - \epsilon} \approx 1 - \epsilon/2$,[5] which is valid for small ϵ, this length becomes $R\sqrt{1 - \ell^2/R^2} \approx R(1 - \ell^2/2R^2) = R - \ell^2/2R$. The length BD is therefore $R - (AB) = \ell^2/2R$.

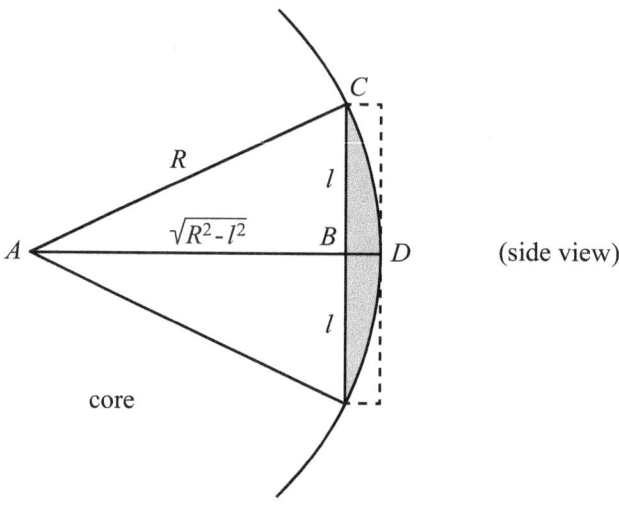

Figure 3.7

As a first approximation in finding the volume of the eaten part, let's use the area of the dashed rectangle in Fig. 3.7, instead of the shaded area. This rectangle has area $(\ell^2/2R)(2\ell) = \ell^3/R$. The curvature of the thin band representing the eaten part (wrapped around the apple along the equator) is very small in the limit we're dealing with, so we can unwrap it with negligible error, giving us a long straight parallelepiped with length $2\pi R$ and cross-sectional area ℓ^3/R. The volume is therefore $(2\pi R)(\ell^3/R) = 2\pi \ell^3 = 2\pi(h/2)^3 = \pi h^3/4$. This is $3/2$ times the correct answer of $\pi h^3/6$ we found earlier. The missing factor of $2/3$ comes from the fact that the area of the shaded region in Fig. 3.7 is $2/3$ times the area of the dashed rectangle. This follows from the facts that (as you can show as an exercise) the curved boundary of the shaded region is essentially a parabola, and the area below (or to the right of, here) a parabola is always $1/3$ of the area of the bounding rectangle, which means that the area above (or to the left of, here) a parabola is $2/3$ of the area of the bounding rectangle.

In most cases, solving a problem in a certain limit is quicker than solving it exactly. Although this wasn't the case in the wide-core limit here, you should never pass up a chance to solve a problem (or at least a certain limit of it) in a second way, as a double check on your first solution! ♣

---

[5]Even if you haven't seen Taylor series before, you can just plug some small ϵ's into a calculator and verify that this relation is (essentially) valid. Alternatively, squaring both sides gives $1 - \epsilon \approx 1 - \epsilon + \epsilon^2/4$, which is valid to first order in ϵ; this is a perfectly valid method for deriving this Taylor series, even though the standard method involves calculus. See the appendix for a review of Taylor series.

10. **Viewing the spokes**

The wheel's contact point on the ground doesn't look blurred, because it is instantaneously at rest (assuming the wheel isn't slipping).[6] But although this is the only point on the wheel that is at rest, there will be other locations in the picture where the spokes do not appear blurred.

The relevant property of a point in the picture where a spoke does not appear blurred is that the point lies on the spoke throughout the duration of the camera's exposure. (The point in the picture need not, however, actually correspond to the same atom on the spoke.) At a given instant, consider a spoke in the lower half of the wheel. A short time later, the spoke will have moved (via both translation and rotation), but it will intersect its original position. The spoke will not appear blurred at this intersection point, because at this point the spoke has no motion perpendicular to itself, which would cause the thickness of the spoke's image in the picture to increase; that's what blurriness is. We must therefore find the locus of these intersections. We can do this in two ways.

FIRST METHOD: Let R be the radius of the wheel. Consider a spoke that makes an angle of θ with the vertical at a given instant. If the wheel then rolls rightward through a small angle $d\theta$, the center moves a distance $R\,d\theta$ rightward. (This follows from the non-slipping assumption, as you can show.) The spoke's motion is a combination of a translation through this distance $R\,d\theta$, plus a clockwise rotation through the angle $d\theta$ (around the top end, at the center). Let r be the radial position of the intersection of the initial and final positions of the spoke, as shown in Fig. 3.8. We can determine r by writing down two expressions for the short segment drawn perpendicular to the initial position of the spoke.

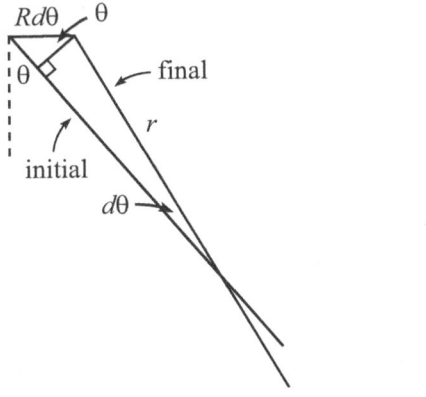

Figure 3.8

---

[6]The wheel's contact point is indeed instantaneously at rest, because for this specific point, the forward motion of the wheel's center is canceled by the backward motion relative to the center, due to the rotation. If the contact point *weren't* instantaneously at rest, then the air near any roadway would be filled with the sound, smoke, and smell of screeching tires. And we would need to change our tires every day (or actually, more like every few minutes).

The two expressions are $(R\,d\theta)\cos\theta$ (from looking at the little right triangle at the top of the figure), and $r\,d\theta$ (to first order in $d\theta$, by looking at the long thin right triangle). Equating these gives $r = R\cos\theta$. This describes a circle whose diameter is the lower vertical radius of the wheel, as shown in Fig. 3.9. This is true because for any θ, the $r = R\cos\theta$ length is one leg of a right triangle whose 90° angle always subtends the 180° arc of half the circle. (An inscribed angle in a circle equals half the arc subtended.) This circle is therefore the locus of points where the spokes don't appear blurred.

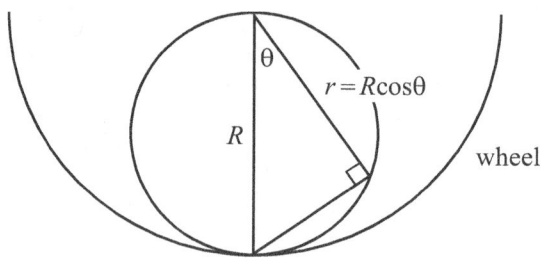

Figure 3.9

There are no non-blurred points in the upper half of the wheel, because the combination of the translation and rotation motions causes each spoke to never intersect its previous position.

SECOND METHOD: Since the wheel's contact point with the ground is instantaneously at rest, the wheel may be considered to be instantaneously rotating around this point. This means that every atom in the wheel (both in the spokes and the rim) instantaneously traces out a tiny arc of a circle centered at the contact point. These circles are shown for two points, P_1 and P_2, in Fig. 3.10; Q is the contact point. A spoke will not appear blurred at the point where this circular motion is along the direction of the spoke. That is, a spoke will not appear blurred at the point where the dashed circle is tangent to the spoke, as is the case for point P_2. At P_2, the spoke's motion is along itself, which means that this point in the picture lies on the spoke throughout the duration of the camera's exposure. At any other point, such as P_1, the spoke's motion is not along itself; there is a component of the motion that is perpendicular to the spoke, which makes the image blurry.

We are therefore concerned with the locus of points P such that the segments PQ and PO are perpendicular. As seen above in Fig. 3.9, this locus is the circle whose diameter is the lower vertical radius of the wheel.

If you want to test this result by actually taking a picture (with a stationary camera and non-negligible exposure time) of a rolling bicycle wheel, make sure you're looking at the front wheel of a bike and not the back one! The front wheel's spokes are radial, whereas the back wheel's spokes are slightly non-radial. (You can ponder why.)

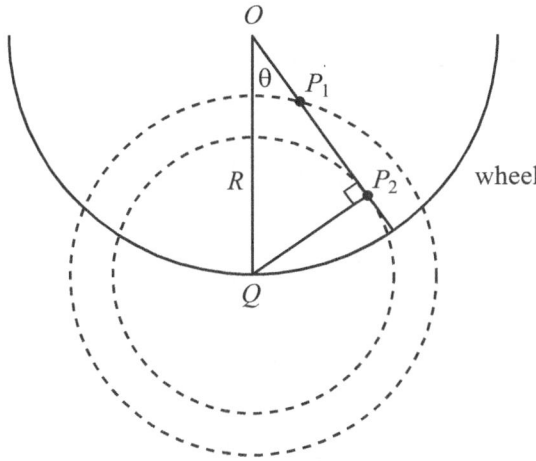

Figure 3.10

11. Painting a funnel

It is true that the volume of the funnel is finite, and that you can fill it up with paint. It is also true that the surface area is infinite, but you actually *can* paint it.

The apparent paradox arises from essentially comparing apples and oranges. In our case we are comparing *volumes* (which are three dimensional) with *areas* (which are two dimensional). When someone says that the funnel can't be painted, he is saying that it would take an infinite *volume* of paint to cover it. But the fact that the surface *area* is infinite does *not* imply that it takes an infinite *volume* of paint to cover it. To be sure, if we tried to paint the funnel with a given fixed thickness of paint, then we would indeed need an infinite volume of paint. But in this case, if we looked at very large values of x where the funnel has negligible cross-sectional thickness, we would essentially have a tube of paint with a fixed radius (the thickness of the paint layer), extending to $x = \infty$, with the funnel taking up a negligible volume at the center of the tube. This tube certainly has an infinite volume.

But what if we paint the funnel with a decreasing thickness of paint, as x gets larger? For example, if we make the thickness be proportional to $1/x$, then the volume of paint is proportional to $\int_1^\infty (1/x)(1/x)\,dx$, which is finite. (The first $1/x$ factor here comes from the $2\pi r$ term in the area, and the second $1/x$ factor comes from the thickness of the paint. We have ignored the $\sqrt{1 + y'^2}$ factor, which goes to 1 for large x.) In this manner, we can indeed paint the funnel. To sum up, you buy paint by the gallon (a volume), not by the square foot (an area). And a gallon of paint can cover an infinite area, as long as you make the thickness go to zero fast enough. The moral of this problem, therefore, is to not mix up things (like volume and area) that have different units/dimensions!

12. Tower of circles

Let the bottom circle have radius 1, and let the second circle have radius r. From Fig. 3.11 we have

$$\sin \beta = \frac{1-r}{1+r}, \qquad \text{where } \beta \equiv \alpha/2. \tag{3.27}$$

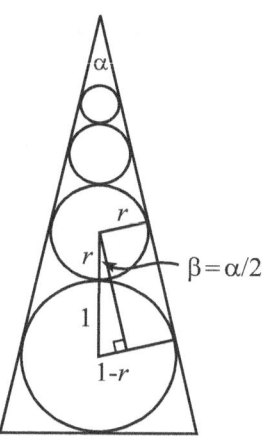

Figure 3.11

In solving this problem, it is easier to work with r instead of the angle α. So we will find the value of r for which A_C/A_T is maximum and then use Eq. (3.27) to obtain α. Note that r is the ratio of the radii of any two adjacent circles. This follows from the fact that we could have drawn the thin right triangle in Fig. 3.11 by using any two adjacent circles. The circles therefore have radii of $1, r, r^2, r^3$, etc.

The area A_T of the triangle can be calculated in terms of r and N as follows. Since we could imagine stacking an infinite number of circles up to the vertex of the triangle, we see that the height h of the triangle is given by the infinite geometric series,

$$h = 2 + 2r + 2r^2 + 2r^3 + \cdots = \frac{2}{1-r}. \tag{3.28}$$

If b is the length of the base of the triangle, then $b/2 = h \tan \beta$. And from the value of $\sin \beta$ given in Eq. (3.27), you can quickly show that $\tan \beta = (1-r)/(2\sqrt{r})$. Therefore,

$$b = 2h \tan \beta = 2 \cdot \frac{2}{1-r} \cdot \frac{1-r}{2\sqrt{r}} = \frac{2}{\sqrt{r}}. \tag{3.29}$$

The area of the triangle is then

$$A_T = \frac{bh}{2} = \frac{1}{2} \cdot \frac{2}{\sqrt{r}} \cdot \frac{2}{1-r} = \frac{2}{\sqrt{r}(1-r)}. \tag{3.30}$$

The total area of the N circles is the geometric series,

$$A_C = \pi\left(1 + r^2 + r^4 + \cdots r^{2(N-1)}\right)$$
$$= \pi\frac{1 - r^{2N}}{1 - r^2}. \tag{3.31}$$

Therefore, the ratio of the areas is

$$\frac{A_C}{A_T} = \frac{\pi}{2}\frac{\sqrt{r}(1 - r^{2N})}{1 + r}. \tag{3.32}$$

Setting the derivative of this equal to zero to obtain the maximum, you can show (with some messy algebra) that the result is

$$1 - r - (4N + 1)r^{2N} - (4N - 1)r^{2N+1} = 0. \tag{3.33}$$

If $N = 1$, we obtain the cubic equation $1 - r - 5r^2 - 3r^3 = 0$. This fortunately has the easily guessable root of -1 (a double root, in fact).[7] You can then show that the remaining root (the one we're concerned with) is $1/3$. From Eq. (3.27), $r = 1/3$ corresponds to $\beta = 30° \implies \alpha = 60°$. So we have an equilateral triangle, which is probably what you would expect for one circle. With $r = 1/3$, Eq. (3.32) gives $(A_C/A_T)_{max} = \pi/3\sqrt{3} \approx 0.60$, which you can work out from scratch for one circle, if you wish.

Note that the above derivation of Eq. (3.33) is valid even though we have only one circle in the $N = 1$ case. In drawing the thin right triangle in Fig. 3.11, we could imagine drawing a second circle above the given one, even though the second circle doesn't actually exist. The above expressions for the various areas in terms of r are still valid.

For general values of N, Eq. (3.33) can only be solved for r numerically.[8] However, if N is large, we can obtain an approximate solution. To leading order in N, we may set $4N \pm 1 \approx 4N$. We may also set $r^{2N+1} \approx r^{2N}$, because r must be very close to 1, otherwise there would be nothing to cancel the "1" term in Eq. (3.33). For convenience, let us write $r \equiv 1 - \epsilon$, where ϵ is very small. Eq. (3.33) then gives

$$(1 - r) - 8Nr^{2N} \approx 0 \implies 8N(1 - \epsilon)^{2N} \approx \epsilon. \tag{3.34}$$

But $(1-\epsilon)^{2N} \approx e^{-2N\epsilon}$. (This follows from Eq. (1.5) in Problem 53. The condition under which it is valid is $\epsilon \ll 1/\sqrt{N}$, which we will find to be true.) So Eq. (3.34) becomes

$$e^{-2N\epsilon} \approx \frac{\epsilon}{8N}. \tag{3.35}$$

---

[7]It turns out that -1 is a double root of Eq. (3.33) for *any* (integral) value of N. To demonstrate this, you can factor out a $(1 + r)$ and then show that $r = -1$ makes the resulting polynomial equal to zero. Or you can show that $r = -1$ makes both the lefthand side of Eq. (3.33) and its derivative equal to zero.

[8]In the $N = 2$ case, the quintic equation reduces to a cubic after the double roots of -1 are taken into account. So this case can still be solved exactly. But it's much easier to just solve it numerically anyway!

Taking the log of both sides gives

$$\epsilon \approx \frac{1}{2N} \ln\left(\frac{8N}{\epsilon}\right)$$

$$\approx \frac{1}{2N} \ln\left(\frac{8N}{\frac{1}{2N}\ln\left(\frac{8N}{\epsilon}\right)}\right), \quad \text{etc.} \tag{3.36}$$

Therefore, to leading order in N, we have (with O shorthand for "of order")

$$\epsilon \approx \frac{1}{2N}\ln\left(\frac{16N^2}{O(\ln N)}\right) = \frac{\ln 16 + 2\ln N - O(\ln \ln N) + \cdots}{2N} \approx \frac{\ln N}{N}. \tag{3.37}$$

Note that for large N, this result for ϵ is much less than $1/\sqrt{N}$, so Eq. (3.35) is indeed valid. Since

$$r \equiv 1 - \epsilon \approx 1 - \frac{\ln N}{N}, \tag{3.38}$$

we can use Eq. (3.27) to obtain

$$\alpha = 2\beta \approx 2\sin\beta = 2\cdot\frac{1-r}{1+r} \approx 2\cdot\frac{\epsilon}{2} = \epsilon, \tag{3.39}$$

where we have used the small-angle approximation $\sin\beta \approx \beta$ (measured in radians). So

$$\alpha \approx \frac{\ln N}{N}, \tag{3.40}$$

measured in radians. This is the desired expression for α, to leading order in N. By leading order, we mean that as N becomes very large, this answer for α becomes arbitrarily close (multiplicatively) to the true answer.

REMARKS:

1. The radius R_N of the top circle in the stack is r^{N-1}. For large N, this equals

$$R_N = r^{N-1} \approx r^N = (1-\epsilon)^N. \tag{3.41}$$

Using Eq. (3.34) and then Eq. (3.37), we have

$$R_N \approx \sqrt{\frac{\epsilon}{8N}} \approx \frac{\sqrt{\ln N}}{2\sqrt{2}\,N}. \tag{3.42}$$

2. The distance from the center of the top circle to the vertex is $R_N/\sin\beta$. For large N, this equals

$$\frac{R_N}{\sin\beta} \approx \frac{R_N}{\beta} = \frac{R_N}{\alpha/2} \approx \frac{\frac{\sqrt{\ln N}}{2\sqrt{2}\,N}}{\frac{1}{2}\cdot\frac{\ln N}{N}} = \frac{1}{\sqrt{2\ln N}}. \tag{3.43}$$

This goes to zero (very slowly) for large N.

3. Since $r \approx 1 - (\ln N)/N$ for large N, the r and $\sqrt{r}$ terms in Eq. (3.32) are essentially equal to 1, and the r^{2N} term is essentially equal to zero. (From the first equation in Eq. (3.34), r^{2N} equals $(1-r)/8N$, which in turn is approximately equal to $(\ln N)/8N^2$.) So we obtain $A_C/A_T \approx \pi/4$. This is the expected answer, because if we look at a small number of adjacent circles, they appear to be circles inside a rectangle (because the long sides of the isosceles triangle are nearly parallel for small α), and you can quickly show that $\pi/4$ is the answer for the rectangular case. Each circle is effectively inside a square.

4. Using Eq. (3.32), along with $r = (1-\sin\beta)/(1+\sin\beta)$ from Eq. (3.27), we can make a plot of $(4/\pi)(A_C/A_T)$ as a function of $\sin\beta$. Fig. 3.12 shows the plot for $N = 10$. In the $N \to \infty$ limit, the left part of the curve approaches a vertical segment, and the rest of the curve approaches a quarter circle. That is, $(4/\pi)(A_C/A_T) \approx \sqrt{1-\beta^2} = \cos\beta$, for $N \to \infty$. The intuitive reason for this is the following. If N is large, and if β is larger than order $(1/N)\ln N$, then we effectively have an infinite number of circles in the triangle (in the sense that they essentially go right up to the vertex). In this infinite case, the ratio A_C/A_T is given by the ratio of the area of a circle to the area of a circumscribing trapezoid whose sides are tilted at an angle β with respect to the vertical. As an exercise, you can show that this ratio is $(\pi/4)\cos\beta$.

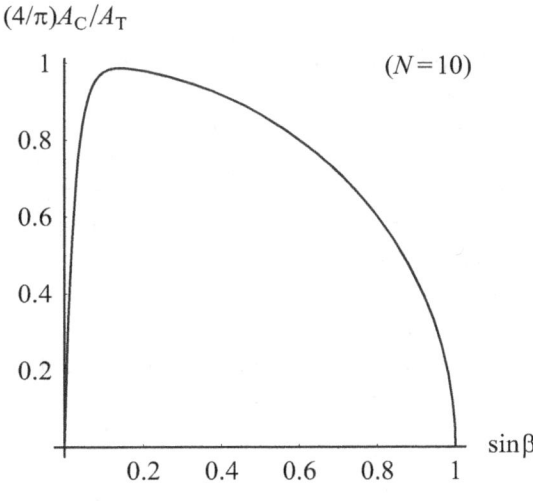

Figure 3.12

5. Eq. (3.33) can be solved numerically for r, for any value of N. Once we've found the r that maximizes A_C/A_T, we can find the corresponding α from Eq. (3.27) and the maximum A_C/A_T value from Eq. (3.32). A few results are shown in Table 3.5.

N	r	α (deg)	α (rad)	$(\ln N)/N$	$(4/\pi)(A_C/A_T)$
1	0.333	60	1.05	0	0.770
2	0.459	43.6	0.760	0.347	0.887
3	0.539	34.9	0.609	0.366	0.931
10	0.754	16.1	0.282	0.230	0.987
100	0.953	2.78	0.0485	0.0461	0.999645
1000	0.9930	0.400	$6.98 \cdot 10^{-3}$	$6.91 \cdot 10^{-3}$	$1 - 6.96 \cdot 10^{-6}$
10^6	0.9999864	$7.76 \cdot 10^{-4}$	$1.36 \cdot 10^{-5}$	$1.38 \cdot 10^{-5}$	$1 - 2.47 \cdot 10^{-11}$

Table 3.5

For large N, we found above that

$$\alpha \text{ (rad)} \approx \frac{\ln N}{N} \quad \text{and} \quad r \approx 1 - \frac{\ln N}{N}. \tag{3.44}$$

These approximate expressions agree well with the numerical results, for large N. Also, for $N = 10$, the $r = 0.754$ value in the table gives $\sin \beta = 0.14$ from Eq. (3.27), which is consistent with a visual inspection of the location of the maximum in Fig. 3.12. ♣

13. **Ladder envelope**

    Assume that the ladder has length 1, for simplicity. In Fig. 3.13, let the ladder slide from segment AB to segment CD. Let CD make an angle θ with the floor, and let AB make an angle $\theta + d\theta$, with $d\theta$ small. The given problem is equivalent to finding the locus of intersections, P, of adjacent ladder positions AB and CD. These are the points where the sliding ladder has no motion perpendicular to itself. This is a necessary property of any point on the envelope, because otherwise the ladder would sweep out area on both sides of the given point, contradicting the fact that the ladder always lies on one side of the envelope.

    Put the ladder in a coordinate system with the floor as the x-axis and the wall as the y-axis. Let a vertical line through B intersect CD at point E. We will find the x and y coordinates of point P by determining the ratio of similar triangles ACP and BEP. We will find this ratio by determining the ratio of AC to BE. AC is given by

    $$AC = \sin(\theta + d\theta) - \sin\theta \approx \cos\theta \, d\theta. \tag{3.45}$$

    This $\cos\theta \, d\theta$ result follows from the fact that if we divide both sides by $d\theta$, the lefthand side is the definition of the derivative of $\sin\theta$, which we know is $\cos\theta$. Alternatively, you can use the trig sum formula, $\sin(\theta + d\theta) = \sin\theta\cos d\theta + \cos\theta\sin d\theta$, and then note that for small $d\theta$ we have $\cos d\theta \approx 1$ and $\sin d\theta \approx d\theta$. (These are the first terms in the Taylor series for $\cos x$ and $\sin x$; see the appendix for a review of Taylor series.) By similar reasoning, we have

    $$BD = \cos\theta - \cos(\theta + d\theta) \approx \sin\theta \, d\theta, \tag{3.46}$$

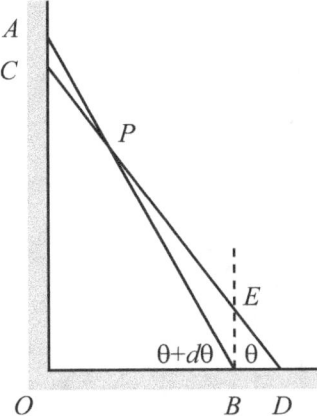

Figure 3.13

with the righthand side being the negative of the derivative of $\cos\theta$ (multiplied by $d\theta$). BE is then given by

$$BE = BD\tan\theta \approx \tan\theta \sin\theta\, d\theta. \tag{3.47}$$

The ratio of triangle ACP to triangle BEP is therefore

$$\frac{\triangle ACP}{\triangle BEP} = \frac{AC}{BE} \approx \frac{\cos\theta\, d\theta}{\tan\theta \sin\theta\, d\theta} = \frac{\cos^2\theta}{\sin^2\theta} \equiv r. \tag{3.48}$$

Since this is the ratio of the triangles, it is also the ratio of the horizontal distances $(AP)_x$ and $(PB)_x$. The ratio of $(AP)_x$ to the entire distance $(AB)_x = (AP)_x + (PB)_x$ is therefore $r/(r+1)$. For very small $d\theta$, we have $(AB)_x = OB \approx OD = \cos\theta$. So the x coordinate of P is

$$P_x = (AP)_x = \frac{r}{r+1}(AB)_x \approx \frac{\cos^2\theta}{\cos^2\theta + \sin^2\theta}\cos\theta = \cos^3\theta. \tag{3.49}$$

Likewise, you can show that the y coordinate of P is $P_y = \sin^3\theta$. The envelope of the ladder may therefore be described parametrically by

$$(x, y) = (\cos^3\theta, \sin^3\theta), \qquad \pi/2 \geq \theta \geq 0. \tag{3.50}$$

Equivalently, using $\cos^2\theta + \sin^2\theta = 1$, the envelope may be described by the equation,

$$x^{2/3} + y^{2/3} = 1. \tag{3.51}$$

The envelope, along with a number of ladder positions, is shown in Fig. 3.14

As double check on Eq. (3.51), there's an easy point (in addition to the points $(0, 1)$ and $(1, 0)$) that we know lies on the envelope. When the ladder is tilted at $45°$, the horizontal and vertical spans of it are each $1/\sqrt{2}$, which means that the midpoint is located at $(1/2\sqrt{2}, 1/2\sqrt{2}) = (1/2^{3/2}, 1/2^{3/2})$. The midpoint must lie on the envelope (this is fairly clear, but see if you can prove why). And indeed, its coordinates satisfy $x^{2/3} + y^{2/3} = 1/2 + 1/2 = 1$.

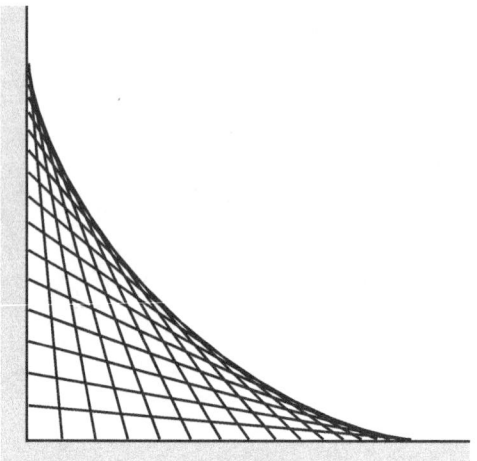

Figure 3.14

14. Equal segments

Let the given line segment be AB, as shown in Fig. 3.15. To begin the process, there are only so many things we can do. Let's pick an arbitrary point P on the side of AB opposite to the given infinite line L (although the other side would work fine too). From P, draw lines through A and B to generate the points M and N on L. The lines AN and BM then generate point Q, and the line PQ generates points C and D.

We claim that C is the midpoint of the given segment AB. This is true because the ratio of similar triangles PCB and PDN is the same as for PCA and PDM (because the ratios of the altitudes from P are the same). So with the lengths indicated in Fig. 3.15, we have $b/d = a/c$. Likewise, the ratio of similar triangles QCB and QDM is the same as for QCA and QDN. So $b/c = a/d$. Multiplying the two preceding equations gives $b^2/dc = a^2/cd \implies b = a$, as desired.

To proceed further and divide AB into three equal segments, there are still only so many lines we can draw, although we have some choices now. We can draw the lines MC and NC, or we can draw the lines AD and BD. Either set will work, but let's pick the former. In any case, the procedure described below shows that if you simply draw every possible line you can draw in 30 seconds, you'll undoubtedly divide AB into three equal segments, whether you know it or not.

At this point, instead of explicitly solving the $N = 3$ case, let's be general and demonstrate inductively how to divide AB into $N + 1$ equal segments, given that we have already divided it into N equal segments. For purposes of concreteness and having a manageable figure, we'll consider the relatively simple case of $N = 3$. It will be clear how to generalize to arbitrary N.

In Fig. 3.16, let the segment AB be divided into three equal segments (the inductive hypothesis) by C_1 and C_2. From an arbitrary point P (assume that P is on the side of AB opposite to the infinite line L, although it need not be), draw

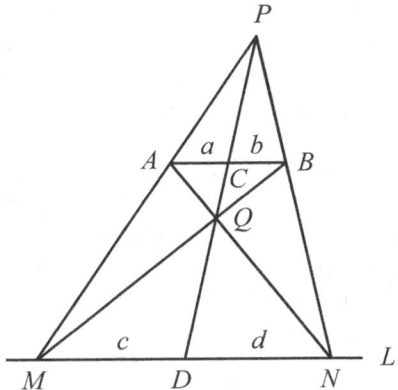

Figure 3.15

lines through A and B to generate the points M and N on L. Draw segments MC_1, MC_2, MB, NA, NC_1, and NC_2. Let the resulting intersections (the ones closest to segment AB) be Q_1, Q_2, and Q_3, as shown.

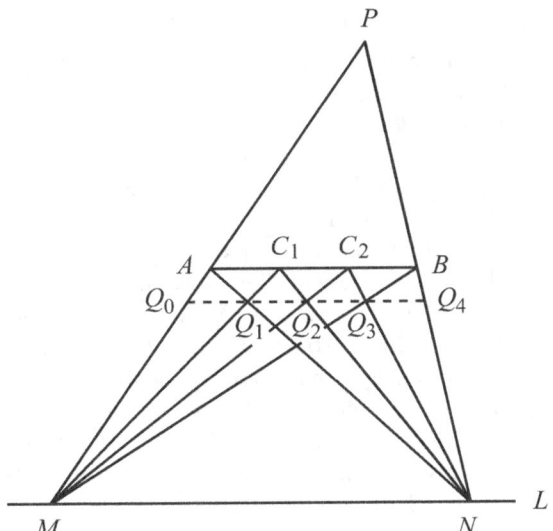

Figure 3.16

Claim: *The intersections of the lines PQ_1, PQ_2, and PQ_3 with AB divide AB into four equal segments.*

Proof: First note that Q_1, Q_2, and Q_3 are collinear on a line parallel to AB (and L). This is true because the ratio of similar triangles AQ_1C_1 and NQ_1M is the same as the ratio of similar triangles $C_1Q_2C_2$ and NQ_2M, because $AC_1 = C_1C_2$.

Therefore, the altitude from Q_1 to AC_1 equals the altitude from Q_2 to C_1C_2. The same reasoning applies to Q_3, so all the Q_i are equal distances from AB. That is, the line determined by the Q_i is parallel to AB (and L). Let this line intersect PM and PN at Q_0 and Q_4, respectively.

We now claim that the distances Q_0Q_1, Q_1Q_2, Q_2Q_3, and Q_3Q_4 are all equal. They are equal because the ratio of similar triangles AQ_0Q_1 and AMN is the same as the ratio of similar triangles $C_1Q_1Q_2$ and C_1MN, because the ratio of the altitudes from A in the first pair is the same as the ratio of the altitudes from C_1 in the second pair. Hence, $Q_0Q_1/MN = Q_1Q_2/MN \implies Q_0Q_1 = Q_1Q_2$. Likewise for the other Q_iQ_{i+1} lengths.

Alternatively, consider triangle MAB. The MC_1 and MC_2 lines divide AB into three equal segments, so they also divide Q_0Q_3 into three equal segments. This is fairly intuitive, but to be rigorous: Since the line of the Q_i's is parallel to AB, the ratio of similar triangles MQ_0Q_1 and MAC_1 is the same as the ratio of similar triangles MQ_1Q_2 and MC_1C_2. Therefore, since $AC_1 = C_1C_2$ by assumption, we have $Q_0Q_1 = Q_1Q_2$. In a similar manner, we obtain $Q_1Q_2 = Q_2Q_3$. So $Q_0Q_1 = Q_1Q_2 = Q_2Q_3$. Analogous reasoning holds with triangle NAB, so all of the Q_iQ_{i+1} lengths are equal.

Therefore, since the Q_i divide Q_0Q_4 into four equal segments, and since Q_0Q_4 is parallel to AB, the intersections of the lines PQ_i with AB divide AB into four equal segments. (To be rigorous, you can follow the line of reasoning in the preceding paragraph.) ■

To divide AB into five equal segments, we can reuse Fig. 3.16, with most of the work having already been done. The only new lines we need to draw are NQ_0 and MQ_4, to give a total of four intersections on a horizontal line one "level" below the Q_i. If we continue with this process, we obtain a figure looking like the one in Fig. 3.17. The horizontal lines in this figure are divided into equal parts by the intersections of the diagonal lines. Lines from the top vertex P down to the dividing points on a given horizontal line divide the original segment into equal parts. The original segment is the top one. This segment therefore serves double duty as both the $N = 1$ and $N = 2$ segments. (The starting procedure in Fig. 3.15 yielded only one point Q, and this single point doesn't determine a line parallel to AB and L. So this step doesn't generate an $N = 2$ line.)

EXTENSION: You are given a line segment with length ℓ, a line parallel to it, and a straightedge. Show how to construct a segment with length $N\ell$, for any integer N. Try to solve this before looking closely at Fig. 3.18, which gives a possible construction. The numbers in the figure represent the order in which things are drawn (shaded numbers for points, unshaded for lines). Prove to yourself why the "9" point yields twice the length of the original segment (the heavy line). Continuing the process rightward yields higher values of N. Note that by combining this extension with the original problem, we can construct a length equal to any rational multiple of ℓ.

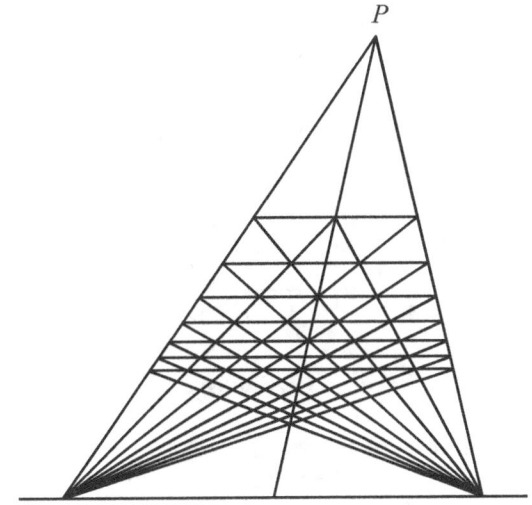

Figure 3.17

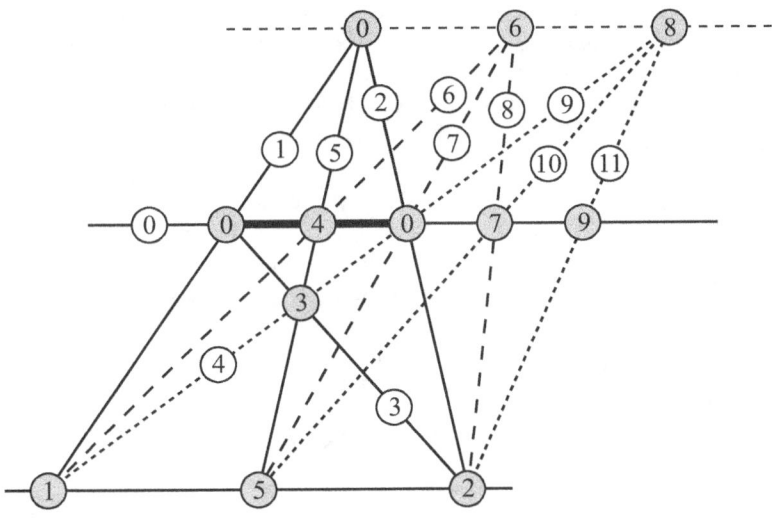

Figure 3.18

15. **Collinear points**

Draw all of the lines determined by the points. From the given assumption in the problem, there are at least three points on each of these lines. Consider all of the distances between any of the points and any of the lines. (Many of these distances are zero, of course, for points lying on a given line.) Assume (in search of a contradiction) that the points don't all lie on a common line, so that some of the distances are nonzero. Since there is a finite number of points and lines, there is a finite number of these distances. Hence there is a smallest nonzero distance, $d_{\min}$ (which may occur more than once). Consider a point P and a line L associated with $d_{\min}$, as shown in Fig. 3.19.

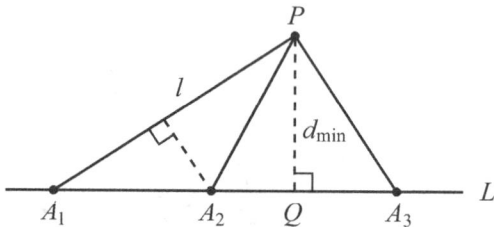

Figure 3.19

Let Q be the projection of P onto L. Since L contains at least three points by assumption, at least two of them must lie on the same side of Q (or one may coincide with Q). Call these points A_1 and A_2, with A_1 being the point farther from Q. Let ℓ be the line through P and A_1. Then the distance from A_2 to ℓ is less than $d_{\min}$ (because this distance is less than or equal to the distance from Q to ℓ, which is strictly less than the distance from Q to P). But this contradicts our assumption that $d_{\min}$ was the smallest nonzero distance from a point to a line. Hence, there can be no smallest nonzero distance. Therefore, all the distances are zero, which means that all the points lie on a common line.

16. **Attracting bugs**

    In all three of the solutions we will give, the key point is that at any time, the bugs form the vertices of a regular N-gon, as shown in Fig. 3.20 for $N = 6$. This is true because this is the only configuration that respects the symmetry of the N bugs. The N-gon will rotate and shrink until it becomes a point at the center.

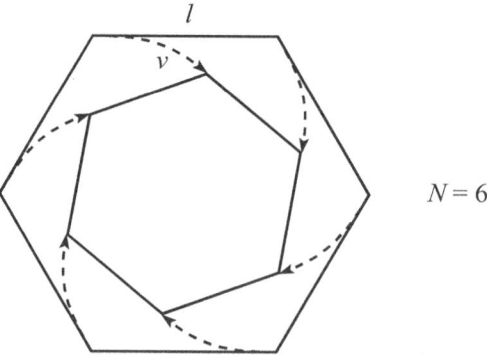

Figure 3.20

FIRST SOLUTION: The important quantity in this solution is the relative speed of two adjacent bugs. By "relative speed" we mean the rate at which the distance between two adjacent bugs decreases. This relative speed is constant, because the relative angle of the bugs' motions is always the same. If the bugs' speed is v, then we see from Fig. 3.21 that the relative speed is $v_r = v(1-\cos\theta)$, where $\theta = 2\pi/N$. Note that the transverse $v\sin\theta$ component of the front bug's velocity is irrelevant here, because it provides no first-order change in the distance between the bugs, for small increments of time dt. From the Pythagorean theorem, it produces only a second-order dt^2 change in the distance.

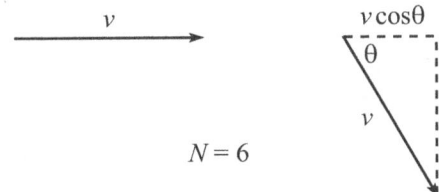

Figure 3.21

For example, if $N = 3$ we have $v_r = 3v/2$; if $N = 4$ we have $v_r = v$; and if $N = 6$ we have $v_r = v/2$. Note also that for $N = 2$ (which doesn't give much of a polygon, being just a straight line) we have $v_r = 2v$, which is correct for two bugs walking directly toward each other. And if $N \to \infty$ we have $v_r \to 0$, which is correct for bugs walking around a circle.

If two bugs start a distance ℓ apart, and if the separation between them decreases at the constant rate of $v(1 - \cos\theta)$, then the time it takes for them to meet is $t = \ell/\bigl(v(1 - \cos\theta)\bigr)$, where $\theta = 2\pi/N$. Therefore, since the bugs walk at speed v, they will each travel a total distance of

$$vt = \frac{\ell}{1 - \cos(2\pi/N)}. \tag{3.52}$$

Note that for a square, this distance equals the length of a side, ℓ. For large N, the Taylor approximation $\cos\theta \approx 1 - \theta^2/2$ gives $vt \approx 2\ell/\theta^2 = N^2\ell/2\pi^2$. (See the appendix for a review of Taylor series.)

The bugs will spiral around an infinite number of times. There are a few ways to see this. First, the future path P_t of the bugs at time t must simply be a scaled-down version of the future path P_0 at the start (because any point in time may be considered to be the start time, with a scaled-down version of the initial separation). This implies that P_t and P_0 must have the same number of spiral revolutions (because scaling down doesn't change the number of revolutions). However, P_0 certainly has more revolutions than P_t, because there is a nonzero rotation between time zero and time t. The number of revolutions in P_0 must therefore be both the same as, and larger than, the number in P_t. The only way this can happen is if the number is infinity. (For example, infinity plus 10 is still infinity.)

Another way of presenting the above reasoning is the following. The number of spiral revolutions, n, cannot be a function of the side length ℓ, because n is dimensionless, whereas ℓ has dimensions of length (and there are no other such parameters available to cancel the length). Therefore, n can depend at most on N. So for a given N, we see that n is fixed. That is, a big N-gon and a small N-gon must have the same n. But it takes time for the big N-gon to become the small N-gon, during which it rotates by some amount. So we reach the same conclusion as above, that the two n's must be both the same and different. Both n's are therefore infinite.

Another line of reasoning is this: We know that after each revolution, a bug's distance from the center decreases by a factor of a, for some fixed a. This a is independent of which revolution the bug is on, due to the scale invariance of the motion.[9] (We will see in the third solution below that $a = e^{-2\pi \tan(\pi/N)}$.) So after n revolutions, the distance from the center decreases by the factor a^n. This becomes zero only for $n = \infty$.

The one exception to this $n = \infty$ result is when $N = 2$, where the bugs simply walk right toward each other, yielding zero revolutions. The escape from the first two equivalent reasonings above is that a larger 2-gon does *not* imply a larger number of revolutions, because the number is zero in all cases. The escape from the last reasoning above is that $a = 0$ when $N = 2$.

SECOND SOLUTION: In this solution, we will determine how quickly the bugs approach the center of the N-gon. A bug's velocity may be decomposed into radial and tangential components, v_R and v_T, as shown in Fig. 3.22. Because at any instant the bugs all lie on the vertices of a regular N-gon, they always walk at the same angle relative to circular motion. Therefore, the magnitudes of v_R and v_T remain constant.

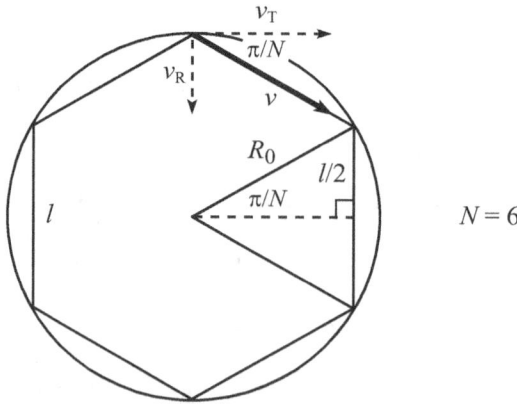

Figure 3.22

---

[9]Of course, bugs of nonzero size would hit each other before they reach the center. If the bugs happen to be very small, then they would eventually require arbitrarily large friction with the floor, in order to provide the centripetal acceleration needed to keep them in a spiral with a very small radius of curvature.

What is the radial component, v_R, in terms of v? The angle between a bug's motion and circular motion is π/N (you can trace this back to the $2\pi/N$ angle subtended by each side), so we have

$$v_R = v \sin(\pi/N). \tag{3.53}$$

What is the radius, R_0, of the initial N-gon? The right triangle in Fig. 3.22 gives us

$$R_0 = \frac{\ell/2}{\sin(\pi/N)}. \tag{3.54}$$

The time for a bug to reach the center is therefore $t = R_0/v_R = \ell/\bigl(2v\sin^2(\pi/N)\bigr)$. So each bug travels a total distance of

$$vt = \frac{\ell}{2\sin^2(\pi/N)}. \tag{3.55}$$

This agrees with Eq. (3.52) in the first solution, due to the half-angle formula, $\sin^2(\theta/2) = (1 - \cos\theta)/2$. The same reasoning used in the first solution shows that the bugs spiral around an infinite number of times.

THIRD SOLUTION: In this solution, we will parameterize a bug's path and then integrate the differential arclength. Let us find a bug's distance, $R(\phi)$, from the center, as a function of the angle ϕ through which it has traveled. The angle between a bug's motion and circular motion is π/N. Therefore, the change in radius, dR, divided by the change in arclength along the circle, $R\,d\phi$, is $dR/(R\,d\phi) = -\tan(\pi/N)$. Separating variables and integrating gives (putting primes on the integration variables)

$$\int_{R_0}^{R} \frac{dR'}{R'} = -\int_0^{\phi} \tan(\pi/N)\,d\phi'$$
$$\implies \ln(R/R_0) = -\phi \tan(\pi/N)$$
$$\implies R(\phi) = R_0\, e^{-\phi \tan(\pi/N)}, \tag{3.56}$$

where R_0 is the initial distance from the center, equal to $\ell/(2\sin(\pi/N))$ from Eq. (3.54) in the second solution. We now see, as we stated in the first solution, that in one revolution (that is, $\Delta\phi = 2\pi$), R decreases by the factor $e^{-2\pi \tan(\pi/N)}$, and that an infinite number of revolutions is required for R to reach zero.

Having found $R(\phi)$, we can integrate the arclength to find the total distance traveled. From the Pythagorean theorem, a little piece of the path has arclength $\sqrt{(R\,d\phi)^2 + (dR)^2}$. The total arclength is therefore (using $dR/d\phi = $

$-R_0 \tan(\pi/N) e^{-\phi \tan(\pi/N)}$ and $1 + \tan^2 x = 1/\cos^2 x$)

$$\int \sqrt{(R\,d\phi)^2 + (dR)^2} = \int_0^\infty \sqrt{R^2 + (dR/d\phi)^2}\, d\phi$$

$$= \int_0^\infty \frac{R_0\, e^{-\phi \tan(\pi/N)}}{\cos(\pi/N)}\, d\phi$$

$$= R_0 \cdot \frac{1}{\cos(\pi/N)} \cdot \frac{-1}{\tan(\pi/N)}\, e^{-\phi \tan(\pi/N)}\bigg|_0^\infty$$

$$= \frac{\ell}{2\sin(\pi/N)} \cdot \frac{1}{\cos(\pi/N)} \cdot \frac{-1}{\tan(\pi/N)} \cdot (-1)$$

$$= \frac{\ell}{2\sin^2(\pi/N)}, \qquad (3.57)$$

in agreement with Eq. (3.55).

REMARK: In the first solution, we found that for large N the total distance traveled is approximately $\ell N^2/2\pi^2$. This result can also be found in the following manner. For large N, a bug's motion can be approximated by a sequence of circles, C_n, with radii $R_n = R_0 e^{-n(2\pi)\tan(\pi/N)} \approx R_0 e^{-n(2\pi^2/N)}$, where we have used $\tan x \approx x$ for small x. To leading order in N, the total distance traveled is therefore the sum of the infinite geometric series (using $e^{-x} \approx 1 - x$ for small x, and $R_0 = \ell/(2\sin(\pi/N)) \approx N\ell/2\pi$ since $\sin x \approx x$ for small x),

$$\sum_{n=0}^\infty 2\pi R_n \approx \sum_{n=0}^\infty 2\pi R_0 e^{-n(2\pi^2/N)}$$

$$= \frac{2\pi R_0}{1 - e^{-2\pi^2/N}}$$

$$\approx \frac{2\pi(N\ell/2\pi)}{2\pi^2/N}$$

$$= \frac{N^2 \ell}{2\pi^2} \cdot \clubsuit \qquad (3.58)$$

17. Find the foci

In all of these constructions, we will assume that you know how to perform standard straightedge-and-compass constructions such as constructing a line parallel to a given line, constructing the perpendicular bisector of a segment, etc. As exercises, you should think about how to do these.

ELLIPSE: Let us first find the center of the ellipse. In Fig. 3.23, draw two arbitrary parallel lines that each meet the ellipse at two points. Call these points A_1 and A_2 on one line, B_1 and B_2 on the other. Bisect segments $A_1 A_2$ and $B_1 B_2$ to yield points A and B. Now repeat this construction with two other parallel lines to give two new bisection points C and D.

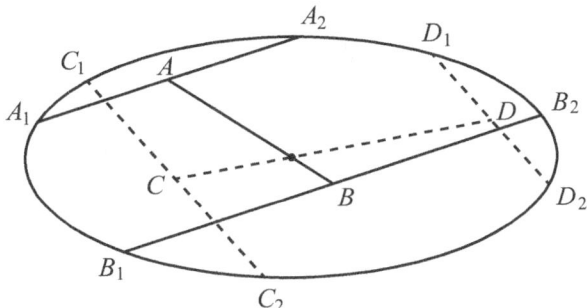

Figure 3.23

Claim 3.1 *The intersection of the line passing through A and B, and the line passing through C and D, is the center of the ellipse.*

Proof: An ellipse is simply a stretched circle (uniform stretching, along one axis), combined perhaps with an overall scaling. And in this uniform stretching process (and overall scaling), all midpoints of segments remain midpoints, straight lines remain straight, and the center of the circle remains the center (of the ellipse now). Therefore, since the line passing through the midpoints of two parallel chords of a circle passes through the center of the circle, the same must be true for an ellipse. The intersection of two such lines is therefore the center of the ellipse. (If this reasoning doesn't satisfy you, we'll give an analytic proof when we get to the hyperbola case.) ∎

Having found the center, we can now find the major and minor axes by drawing a circle, with its center at the center of the ellipse, that meets the ellipse at four points – the vertices of a rectangle. (The radius can be chosen arbitrarily, as long as it lies between the lengths of the semi-minor and semi-major axes, so that the circle does in fact intersect the ellipse at four points.) The axes of the ellipse are the lines parallel to the sides of the rectangle and passing through the center of the ellipse. Equivalently, the axes are the perpendicular bisectors of the sides of the rectangle.

Having found the axes, the foci are the two points on the major axis that are a distance a (where $2a$ is the major-axis length) from the endpoints of the minor axis. (This is true because all points on the ellipse have the property that the sum of the distances to the two foci is $2a$.) These two points can be constructed by using the compass to draw a circle of radius a centered at an endpoint of the minor axis.

PARABOLA: Let us first find the axis of the parabola. In Fig. 3.24 draw two arbitrary parallel lines that each meet the parabola at two points. Call these points A_1 and A_2 on one line, B_1 and B_2 on the other. Bisect segments A_1A_2 and B_1B_2 to yield points A and B.

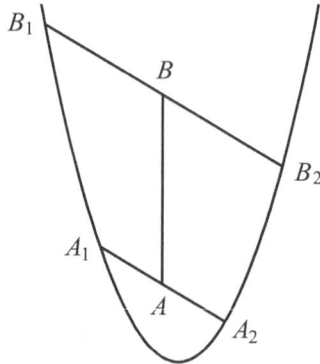

Figure 3.24

Claim 3.2 *Segment AB is parallel to the axis of the parabola.*

Proof: This follows from the reasoning in the ellipse case, along with the fact that a parabola is the limit of an infinitely elongated ellipse, with its center at infinity. From the ellipse reasoning, the extension of *AB* must pass through the center of the ellipse/parabola at infinity, which lies directly (and infinitely far) above the vertex (the bottom point) of the parabola.[10] The only way this can happen is if *AB* is parallel to the axis. (Again, if this reasoning doesn't satisfy you, we'll give an analytic proof when we get to the hyperbola case.) ∎

To obtain the axis of the parabola, draw a line perpendicular to *AB*, which meets the parabola at points *C* and *D*. The perpendicular bisector of *CD* is the axis of the parabola. And then the intersection of the axis and the parabola is the vertex.

Having found the axis, the focus may be found as follows. Call the axis the *y*-axis of a coordinate system, with the parabola opening up in the positive *y*-direction. Let the vertex of the parabola be at $(0, 0)$, and let the focus be at $(0, a)$. Then a horizontal line through the focus meets the parabola at the points $(\pm 2a, a)$. This is true because the distance from each of these points to the focus (which is the absolute value of the *x*-coordinate in this case) must equal the distance from each point to the directrix (which is the horizontal line located a distance *a* below the vertex), which equals $2a$. (This is the definition of a parabola. It is consistent with taking the limit of the definition of an ellipse, with one focus being infinitely far away. You can think about why.) The $(\pm 2a, a)$ points also follow from writing the parabola in the standard form $x^2 = 4ay$, where *a* is the focal distance.

The focus of the parabola may therefore be found by drawing lines through the vertex, with slopes $1/2$ and $-1/2$ (you can think about how). These two lines meet the parabola at points *E* and *F*. The intersection of segment *EF* with the axis is the focus.

---

[10]A parabola can also be thought of as the limit of a hyperbola whose asymptotes are nearly parallel. In this case, the center of the hyperbola/parabola lies infinitely far *below* the vertex in Fig. 3.24.

HYPERBOLA: Let us first find the center of the hyperbola. The same construction works here as did for the ellipse, but we will now present a (lengthier) analytic proof. The following claim is valid for all three types of conic sections.

Claim 3.3 *The center of a conic section is the intersection of two lines, each of which passes through the midpoints of two parallel chords of the conic section.*

Proof: Let the conic section be written as

$$rx^2 + sy^2 = 1. \tag{3.59}$$

This describes an ellipse if r and s are positive, and a hyperbola if r and s have opposite sign. A parabola is obtained in the $r/s \to \pm 0, \pm\infty$ limits (see the remark at the end of the solution). In all cases, the center of the conic section is the origin, $(0,0)$. Our goal is therefore to determine the location of the origin, given the conic section. Consider a line of the form

$$y = ax + b. \tag{3.60}$$

If this line meets the conic section at two points, you can show that the midpoint of the resulting chord has coordinates

$$\left(-\frac{sab}{r+sa^2}, \frac{rb}{r+sa^2}\right). \tag{3.61}$$

Note that when solving the quadratic equation for the intersection of the line and the conic section, you can ignore the discriminant in the quadratic formula, because we are concerned only with the midpoint between (that is, the average of) the intersections. This simplifies things greatly.

The slope of the line joining the above midpoint to the center of the conic section (which is the origin) equals $-r/(sa)$. This is independent of b, so another parallel chord (that is, another chord with the same a but a different y-intercept b in Eq. (3.60)) will also have its midpoint lying on the same line through the origin with slope $-r/(sa)$. Equivalently, the center of the conic section (the origin) lies on the line passing through the midpoints of two parallel chords. The intersection of this line with the analogous line generated by two other parallel chords is therefore the center of the conic section. See Fig. 3.25 for the case of a hyperbola. (The asymptotes have been drawn for clarity, but they aren't relevant to the construction.) ∎

Having found the center of the hyperbola, we can now find the axes by drawing a circle, with its center at the center of the hyperbola, that meets the hyperbola at two points, generating a chord. The axes of the hyperbola are the perpendicular bisector of this chord, along with the line parallel to the chord and passing through the center.

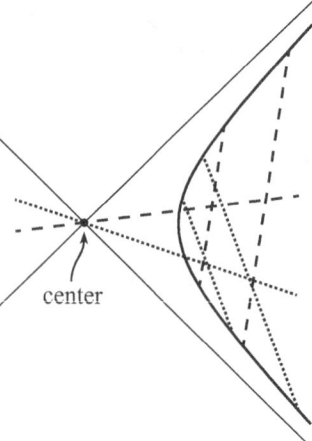

Figure 3.25

Let us now, for convenience, assume that the hyperbola is written in the form,

$$\frac{x^2}{m^2} - \frac{y^2}{n^2} = 1. \tag{3.62}$$

We'll invoke the standard result that the focal length is $c = \sqrt{m^2 + n^2}$. We have already found m, which is the distance from the center to an intersection of the major axis with the hyperbola. So we simply need to find n, which may be found by noting that the point $(\sqrt{2}m, n)$ lies on the hyperbola. We can therefore construct the foci as follows. Knowing the length m, we can construct the length $\sqrt{2}m$ (the diagonal of a square with side m), and then the point $(\sqrt{2}m, 0)$. We can then draw a vertical line to obtain the point $(\sqrt{2}m, n)$ on the hyperbola, which gives us n. Using the known lengths m and n, we can construct a rectangle with these side lengths; the diagonal then has the desired length $\sqrt{m^2 + n^2}$. The foci are the points $(\pm\sqrt{m^2 + n^2}, 0)$. Note that we don't need to be given the other branch of the hyperbola for this construction to work.

REMARK: We mentioned above that a parabola is obtained in the $r/s \to \pm 0, \pm \infty$ limits of the $rx^2 + sy^2 = 1$ equation in Eq. (3.59). This can be shown in various ways; we'll take the following route. (We'll consider the $r/s \to +\infty$ case; the others proceed in a similar manner.) Let $r = \epsilon$ and $s = \epsilon^2$, where $\epsilon \ll 1$. Then $r/s = 1/\epsilon$ is very large. The ellipse given by $\epsilon x^2 + \epsilon^2 y^2 = 1$ is very wide and very tall, but with the height much greater than the width. This is true because x ranges between $\pm 1/\sqrt{\epsilon}$, and y ranges between $\pm 1/\epsilon$. The latter is much larger in the $\epsilon \to 0$ limit.

We claim that the $\epsilon x^2 + \epsilon^2 y^2 = 1$ ellipse looks like a parabola near the top and bottom. We'll demonstrate this for the bottom; the top works out similarly. Solving for y in $\epsilon x^2 + \epsilon^2 y^2 = 1$, and taking the negative square root since we're dealing with the bottom, gives (using the Taylor series $\sqrt{1-z} \approx 1 - z/2$; see the appendix for a review of Taylor series)

$$y = -\frac{1}{\epsilon}\sqrt{1 - \epsilon x^2}$$
$$\approx -\frac{1}{\epsilon}\left(1 - \frac{\epsilon x^2}{2}\right)$$
$$= -\frac{1}{\epsilon} + \frac{x^2}{2}. \tag{3.63}$$

Shifting the origin of our coordinate system downward by letting $y' \equiv y + 1/\epsilon$ yields $y' = x^2/2$. This is the equation for a parabola, as desired. This result is valid as long as the Taylor approximation we used above is valid, which is the case when $\epsilon x^2 \ll 1 \implies x \ll 1/\sqrt{\epsilon}$. So the result is valid (that is, the ellipse looks like a parabola) for x values that are much less than the horizontal span of the ellipse.

Technically, the above ellipse also looks like a parabola near its left and right sides (where it crosses the x-axis). However, it is necessary to rescale the y-axis in an ϵ-dependent manner in order for the parabola to not be infinitely flat in the $\epsilon \to 0$ limit. With the forms we chose above for r and s (namely ϵ and ϵ^2), there was no need for any rescaling. Other forms of r and s, with $r/s \to \pm 0, \pm \infty$, will in general require rescaling. ♣

18. **Construct the center**

    Let the radius of the circle be R (which we don't know yet). Pick an arbitrary point A on the circle, as shown in Fig. 3.26.

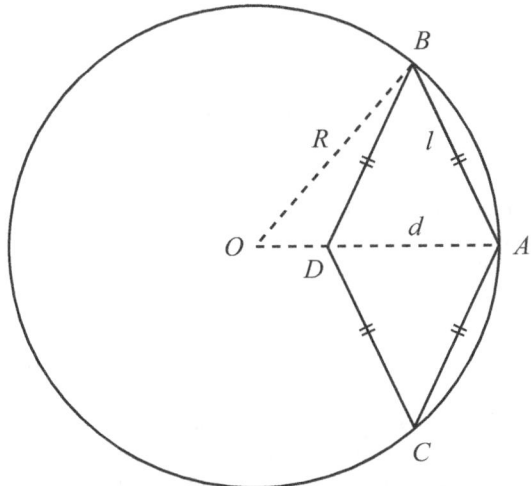

Figure 3.26

Construct points B and C on the circle, with $AB = AC = \ell$, where ℓ is arbitrary. (However, this construction won't work if ℓ is too large or too small. We'll determine these bounds below.) Construct point D with $DB = DC = \ell$. Let the distance DA be d. If O is the location of the center of the circle (which we don't know yet), then triangles AOB and ABD are similar isosceles triangles (because they have $\angle BAD$ in common). Therefore, $OA/BA = BA/DA$, which gives $R/\ell = \ell/d \implies d = \ell^2/R$.

The above construction shows that if we are given a length ℓ and a circle of radius R, then we can construct the length ℓ^2/R. Therefore, we can produce the length R by simply repeating the above construction with the same length ℓ, but now with a circle whose radius is the ℓ^2/R length we just produced. In Fig. 3.27, D is the center of the circle with radius ℓ^2/R (with the circle passing through A). Repeating the construction with this circle gives $GA = \ell^2/(\ell^2/R) = R$. Hence, G is the center of the original circle.

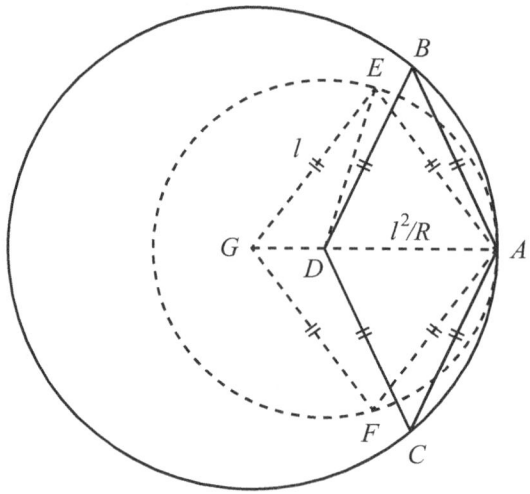

Figure 3.27

If you want to go through the similar-triangles argument, note that triangles ADE and AEG are similar isosceles triangles (because they have $\angle EAD$ in common). Therefore, $DA/EA = EA/GA$, which gives $(\ell^2/R)/\ell = \ell/GA \implies GA = \ell^2/(\ell^2/R) = R$.

RESTRICTIONS: In order for this construction to work, it is necessary (and sufficient) for $R/2 < \ell < 2R$. The upper limit on ℓ comes from the requirement that a circle of radius ℓ (centered at A) intersects the given circle of radius R. (The construction still works even if the intersection points are nearly diametrically opposite to A, as you can verify.) So $\ell < 2R$. The lower limit on ℓ comes from the requirement that a circle of radius ℓ (centered at A) intersects the circle of radius ℓ^2/R (centered at D). This gives $\ell < 2 \cdot \ell^2/R \implies R/2 < \ell$. If you want to write this bound in terms of $\angle BAC$, you can show that $\ell = R/2$ corresponds to $\angle BAC \approx 151°$. So you want $\angle BAC$ to be less than $151°$.

EXTENSION: The above solution can be extended to solve the following problem: Given three points, construct the circle passing through them. You should set the book aside and try to solve this before reading further.

The solution proceeds along the lines of the above solution. In Fig. 3.28, the three given points are A, B, and C. Construct point D with $DB = AB$ ($\equiv \ell_1$) and $DC = AC$ ($\equiv \ell_2$). Let O be the location of the center of the desired circle (which we don't know yet). Then $\angle BOA = \overset{\frown}{BA}$. Also, $\angle DCA = 2(\angle BCA) = 2(\overset{\frown}{BA}/2) = \overset{\frown}{BA}$. Therefore, $\angle BOA = \angle DCA$, and so triangles BOA and DCA are similar isosceles triangles. Hence,

$$\frac{BO}{BA} = \frac{DC}{DA} \implies \frac{R}{\ell_1} = \frac{\ell_2}{DA} \implies DA = \frac{\ell_1 \ell_2}{R}. \qquad (3.64)$$

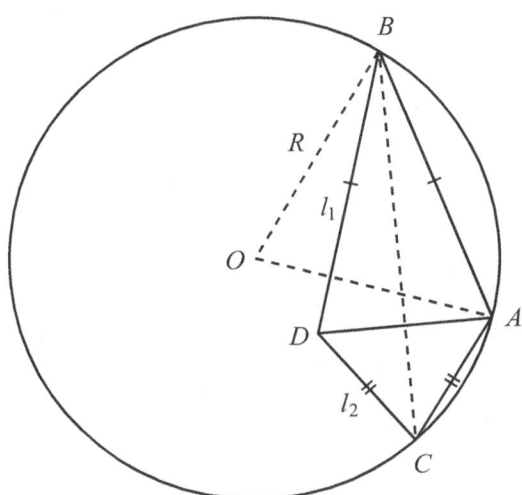

Figure 3.28

As in the above solution, we can apply this construction again, with the same lengths ℓ_1 and ℓ_2, but now with a circle of radius $\ell_1\ell_2/R$ (which we just produced). In Fig. 3.29, D is the center of the circle with radius $\ell_1\ell_2/R$ (with the circle passing through A). We obtain $GA = \ell_1\ell_2/(\ell_1\ell_2/R) = R$. Having found the length R, we can construct the center of the desired circle passing through the given three points. (Just find the intersection of circles with radius R drawn around the points.)

RESTRICTIONS: In order for this construction to work, we must be able to construct points E and F on the circle of radius $\ell_1\ell_2/R$ (centered at D and passing through A). In order for these points to exist, the diameter of this circle must be larger than both ℓ_1 and ℓ_2. That is, $2\ell_1\ell_2/R > \max(\ell_1, \ell_2)$. If $\ell_1 > \ell_2$, this condition becomes $2\ell_1\ell_2/R > \ell_1 \implies \ell_2 > R/2$. Similarly, $\ell_2 > \ell_1$ yields $\ell_1 > R/2$. In either case, we see that the smaller of ℓ_1 and ℓ_2 must be greater than $R/2$. So the condition can be written as $\min(\ell_1, \ell_2) > R/2$. Of course, if we are given three points, then we are given three lengths – the ℓ_1, ℓ_2, and ℓ_3 sides of the triangle the points determine. If two of these lengths are larger than $R/2$, it doesn't matter if

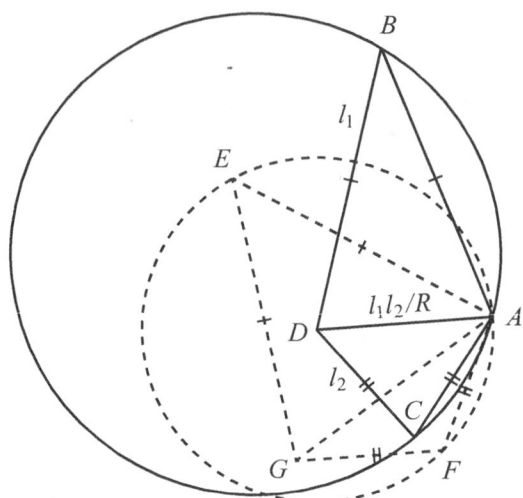

Figure 3.29

the third one isn't. For example, if the angle between AB and AC in Fig. 3.29 were slightly larger (enough to make the radius R of the circle large enough so that $AC < R/2$), then we could just pick BC, instead of AC, as our ℓ_2. However, if we make the angle too large, then we'll end up with $AB < R/2$ too. (For the upper bounds, we need both ℓ_1 and ℓ_2 to be smaller than $2R$, as in the original problem. But this condition is automatically satisfied, because all three points lie on the circle, by assumption.)

What do we do if the lower bound, $\min(\ell_1, \ell_2) > R/2$, isn't satisfied? Simply construct more points on the circle until some three of them satisfy the condition. For example, as shown in Fig. 3.30, construct point B_1 with $B_1 B = CA$ and $B_1 A = CB$. Then triangle $B_1 BA$ is congruent to triangle CAB, so point B_1 also lies on the circle (by symmetry under reflection across the perpendicular bisector of AB). In a similar manner we can construct B_2 as shown, and then B_3, etc., to obtain an arbitrary number of points on the circle. After constructing a sufficient number of points, we will be able to pick three of them that satisfy the condition $\min(\ell_1, \ell_2) > R/2$.

Of course, after constructing these new points on the circle, it is easy to pick three of them that have $\ell_1 = \ell_2$ (for example, B, B_{2n}, and B_{4n}, in the notation of Fig. 3.30). We can then use the easier symmetrical solution in part (a) to find the center of the circle.

19. **Find the angles**

Although this problem seems simple at first glance, angle chasing won't provide the answer. Something a bit more sneaky is required. At the risk of going overboard, we'll present four solutions. You can check that all of the solutions rely on the equality of the two given 50° angles, and on the fact that $2(80°) + 20° = 180°$.

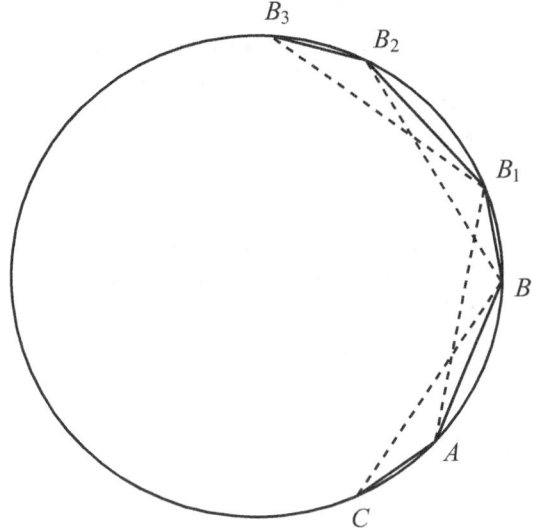

Figure 3.30

FIRST SOLUTION: From the given angles, we find that $\angle ACD = 60°$ and $\angle ABD = 30°$, as shown in Fig. 3.31.

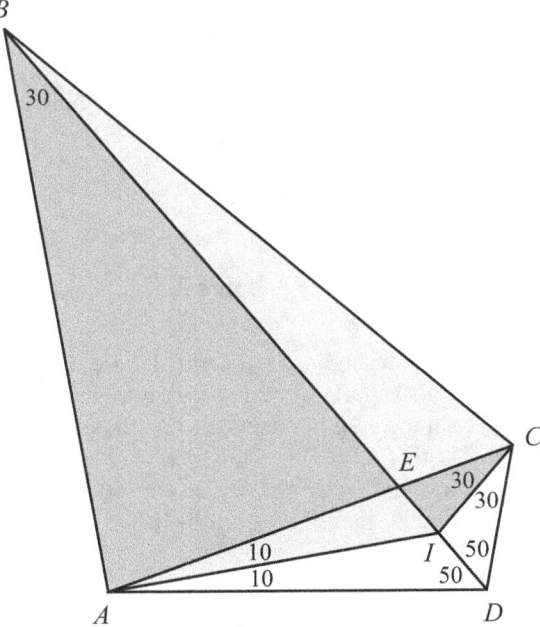

Figure 3.31

Let AC and BD intersect at E. Draw the angle bisectors of triangle ACD. They meet at the incenter, I, located along segment ED. Since $\angle ECI = 30° = \angle EBA$, triangles ECI and EBA are similar (because they also have the common angle at E). Therefore, triangles EBC and EAI are also similar (because they have the same ratio of corresponding sides, along with the common angle at E). Thus, $\angle EBC = \angle EAI = 10°$. We then quickly find $\angle ECB = 60°$.

SECOND SOLUTION: From the given angles, we find that $\angle ABD = 30°$, as shown in Fig. 3.32. Let AC and BD intersect at E. Draw segment AF, with F on BE, such that $\angle EAF = 50°$. We then have $\angle FAB = 30°$. So triangle FAB is isosceles, with $FA = FB$.

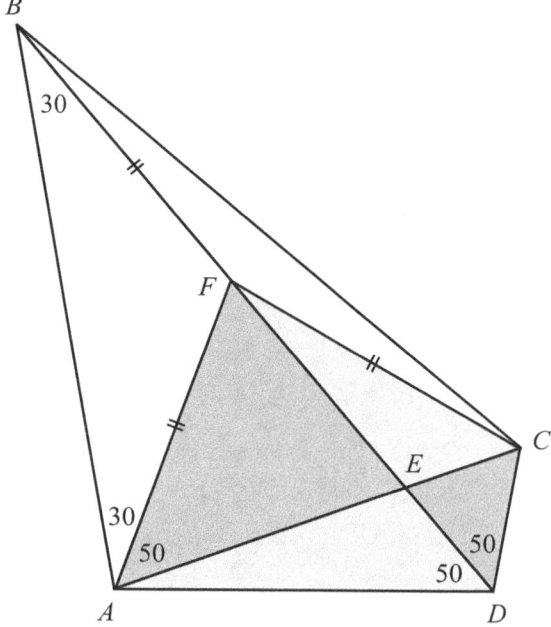

Figure 3.32

Since $\angle EDC = \angle EAF$, triangles EDC and EAF are similar. Therefore, triangles EAD and EFC are also similar (because they have the same ratio of corresponding sides, along with the common angle at E). Hence, $\angle ECF = 50°$, so triangle FCA is isosceles with $FC = FA$. Thus, $FC = FA = FB$, so triangle FBC is also isosceles, with $\angle FBC = \angle FCB$. Since you can quickly show that these two angles must sum to $20°$, they must each be $10°$. Therefore, $\angle FBC = 10°$ and $\angle ECB = 50° + 10° = 60°$.

THIRD SOLUTION: From the given angles, we find that $\angle ACD = 60°$, as shown in Fig. 3.33. Reflect triangle ABC across AB to yield triangle ABG. Note that D, A, and G are collinear because $2(80°) + 20° = 180°$. From the law of sines in triangle DBC, we have

$$\frac{\sin 50°}{BC} = \frac{\sin(60° + \alpha)}{BD}. \tag{3.65}$$

From the law of sines in triangle DBG, we have

$$\frac{\sin 50°}{BG} = \frac{\sin \alpha}{BD}. \tag{3.66}$$

But $BC = BG$, so the preceding two equations yield $\sin(60° + \alpha) = \sin \alpha$. Therefore, $60° + \alpha$ and α are supplementary angles, which gives $\alpha = 60°$. We then quickly obtain $\angle DBC = 10°$.

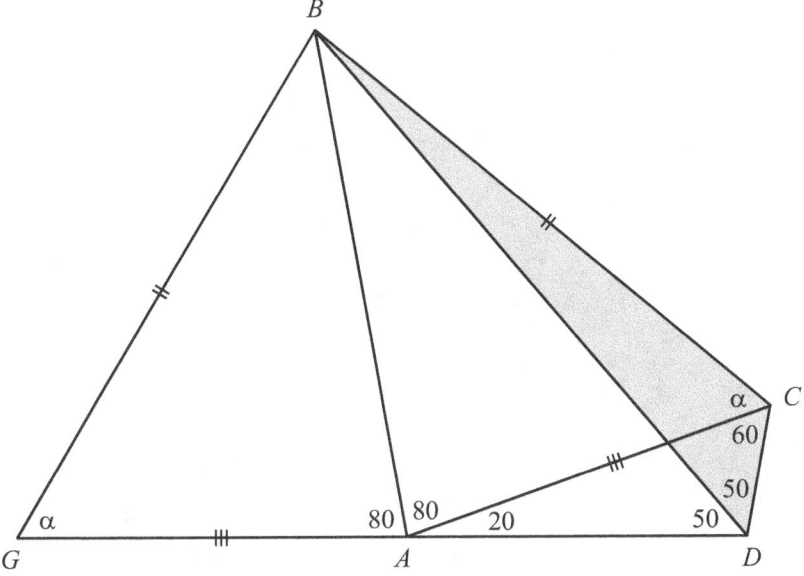

Figure 3.33

FOURTH SOLUTION: We now present the brute-force method using the law of sines, just to show that it can be done. (The law of sines states that $a/\sin A = b/\sin B = c/\sin C$, where A is the angle opposite the side with length a, etc.) The point here is that the four given angles uniquely specify the shape of the quadrilateral, which means that the desired angles are determined. The law of sines allows us to quantify how certain lengths determine others.

In Fig. 3.34, let AC and BD intersect at E. Let the length of AD be 1 unit, and let the lengths a, b, c, and d be as shown. Then the law of sines in triangle AED gives

$$a = \frac{\sin 50°}{\sin 110°} \cdot 1 \quad \text{and} \quad d = \frac{\sin 20°}{\sin 110°} \cdot 1. \tag{3.67}$$

The law of sines in triangles AEB and DEC then gives

$$b = \frac{\sin 80°}{\sin 30°} \cdot \left(\frac{\sin 50°}{\sin 110°}\right) \quad \text{and} \quad c = \frac{\sin 50°}{\sin 60°} \cdot \left(\frac{\sin 20°}{\sin 110°}\right). \quad (3.68)$$

The law of sines in triangle BEC finally gives

$$\left(\frac{\sin 80° \sin 50°}{\sin 30° \sin 110°}\right) \Big/ \sin \alpha = \left(\frac{\sin 50° \sin 20°}{\sin 60° \sin 110°}\right) \Big/ \sin \beta. \quad (3.69)$$

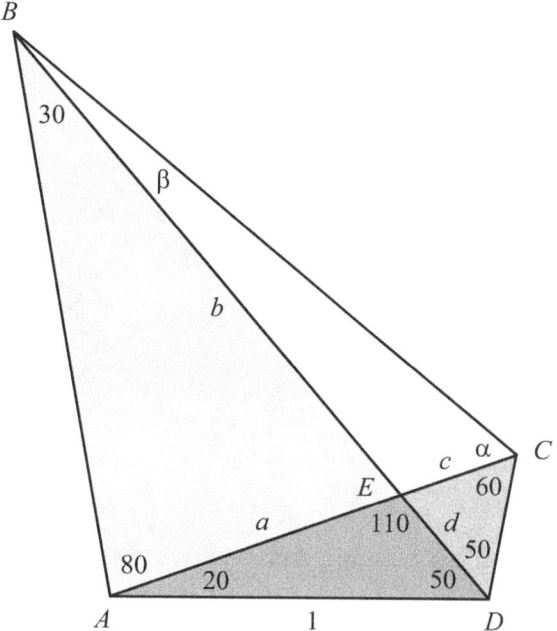

Figure 3.34

Substituting $70° - \alpha$ for β yields (using the trig sum formula for $\sin(x+y)$ and trudging through some algebra)

$$\tan \alpha = \frac{\sin 60° \sin 80° \sin 70°}{\sin 60° \sin 80° \cos 70° + \sin 30° \sin 20°}. \quad (3.70)$$

Using $\sin 20° = 2 \sin 10° \cos 10° = 2 \sin 10° \sin 80°$, along with $\sin 30° = 1/2$, we obtain

$$\tan \alpha = \frac{\sin 60° \sin 70°}{\sin 60° \cos 70° + \sin 10°}. \quad (3.71)$$

Finally, expanding $\sin 10° = \sin(70° - 60°)$ gives the result,

$$\tan\alpha = \tan 60°. \tag{3.72}$$

Hence $\alpha = 60°$, which implies $\beta = 10°$. Not the most elegant solution, but it works!

20. **Rectangle in a circle**

    In Fig. 3.35, let the incenters of triangles ADB and ADC be X and Y, respectively. The incenter of a triangle lies on the angle bisectors, so $\angle XAY$ can be written as

    $$\angle XAY = \angle XAD - \angle YAD = \frac{1}{2}\angle BAD - \frac{1}{2}\angle CAD$$
    $$= \frac{1}{2}\angle BAC = \frac{1}{4}\widehat{BC}. \tag{3.73}$$

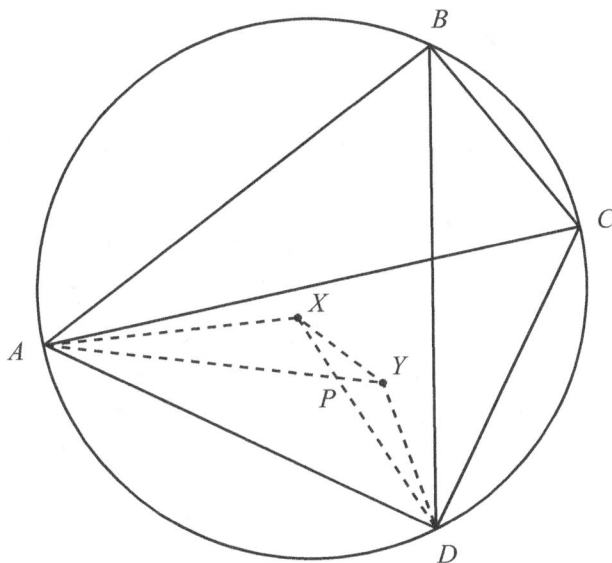

Figure 3.35

A similar argument with $A \leftrightarrow D$, $B \leftrightarrow C$, and $X \leftrightarrow Y$ shows that $\angle YDX$ also equals $(1/4)\widehat{BC}$. This equality of $\angle XAY$ and $\angle YDX$ implies that triangles XAP and YDP are similar (because they also have the common angle at P). Therefore, triangles PXY and PAD are also similar (because they have the same ratio of corresponding sides, along with the common angle at P). Therefore, $\angle PXY = \angle PAD$. The two pairs of equal angles we have just deduced are written as α and β in Fig. 3.36.

We may now repeat the above procedure with the incenters (Y and Z) of triangles DCA and DCB. The result is two more pairs of equal angles, as shown in Fig. 3.37.

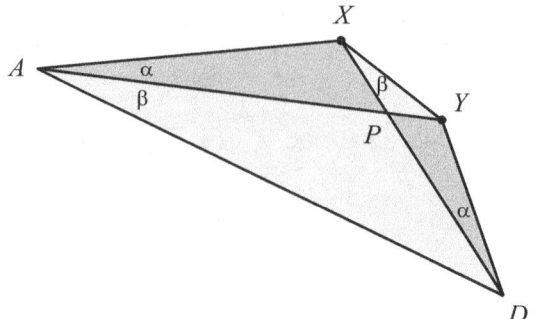

Figure 3.36

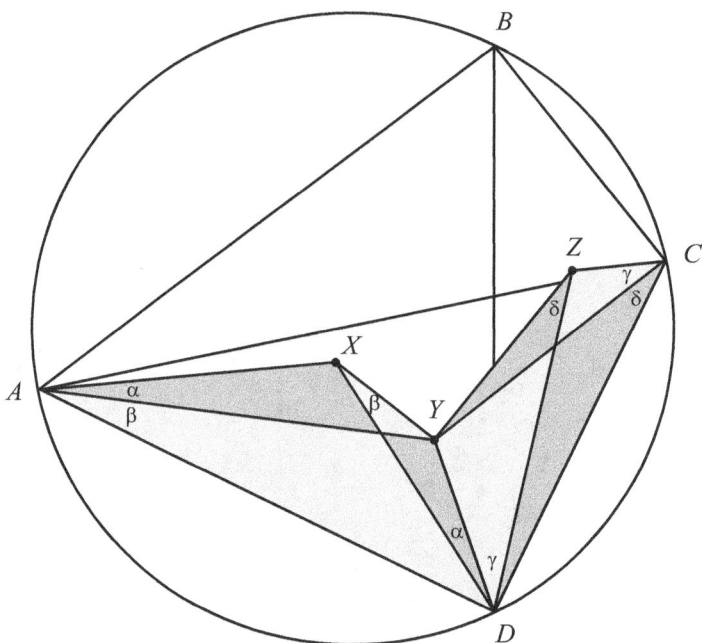

Figure 3.37

The four angles shown have the values,

$$\alpha = (1/4)\,\widehat{BC},$$
$$\beta = (1/2)\angle CAD = (1/4)\,\widehat{CD},$$
$$\gamma = (1/4)\,\widehat{AB},$$
$$\delta = (1/2)\angle ACD = (1/4)\,\widehat{AD}. \tag{3.74}$$

Therefore,
$$\alpha + \beta + \gamma + \delta = \frac{1}{4}(\widehat{BC} + \widehat{CD} + \widehat{AB} + \widehat{AD}) = \frac{1}{4}(360°) = 90°. \quad (3.75)$$

We now note that angle $\angle XYZ$ is given by
$$\begin{aligned}\angle XYZ &= 360° - \angle XYD - \angle ZYD \\ &= 360° - (180° - \alpha - \beta) - (180° - \gamma - \delta) \\ &= \alpha + \beta + \gamma + \delta \\ &= 90°. \end{aligned} \quad (3.76)$$

The same reasoning holds for the three other vertices of the incenter quadrilateral. Therefore, this quadrilateral is a rectangle, as we wanted to show.

21. **Product of lengths**

    (Thanks to Mike Robinson for this solution and generalization.) Put the circle in the complex plane, with its center at the origin. Let the given vertex of the N-gon be located at the point $(1, 0)$. Let $a \equiv e^{2\pi i/N}$, so that $a^N = 1$. Then the other vertices are located at the points a^n, where $n = 1, \ldots, N - 1$. These points are all Nth roots of 1, because $(a^n)^N = (a^N)^n = 1^n = 1$.

    Let the distance between the vertex at $(1, 0)$ and the vertex at a^n be ℓ_n. Then the desired product (call it P_N) of the $N - 1$ segments from the given vertex to the other vertices is

    $$\begin{aligned} P_N &= \ell_1 \ell_2 \ldots \ell_{N-1} \\ &= |1 - a||1 - a^2| \cdots |1 - a^{N-1}| \\ &= (1 - a)(1 - a^2) \cdots (1 - a^{N-1}), \end{aligned} \quad (3.77)$$

    where the third line comes from the fact that the product is real, because $(1 - a^k)$ is the complex conjugate of $(1 - a^{N-k})$, so the phases in the product cancel in pairs.

    Consider the function,
    $$F(z) \equiv z^N - 1. \quad (3.78)$$

    One factorization of $F(z)$ is
    $$F(z) = (z - 1)(z^{N-1} + z^{N-2} + \cdots + 1). \quad (3.79)$$

    Another factorization is
    $$F(z) = (z - 1)(z - a)(z - a^2) \cdots (z - a^{N-1}), \quad (3.80)$$

    because the righthand side is the factorization that yields the zeros of $z^N - 1$, namely all the Nth roots of 1, which are the numbers of the form a^n (including $n = 0$). Equating the above two factorizations and canceling the $z - 1$ factor gives

    $$(z - a)(z - a^2) \cdots (z - a^{N-1}) = z^{N-1} + z^{N-2} + \cdots + 1. \quad (3.81)$$

This equality holds for any value of z. In particular, if we set $z = 1$ the lefthand side becomes the P_N in Eq. (3.77), and the righthand side is just N. So we obtain $P_N = N$, as desired.

If you're worried that Eq. (3.81) might not be valid for $z = 1$ because we derived it by dividing Eqs. (3.79) and (3.80) by $z - 1$, you can just take a limit. Eq. (3.81) certainly holds for z arbitrarily close to 1, and since the functions on the two sides of the equation are continuous, equality must also hold for $z = 1$.

REMARK: Consider the product of the N lengths from an arbitrary point z in the complex plane, to *all* N vertices of the N-gon; see Fig. 3.38. This product equals the absolute value of the righthand side of Eq. (3.80). Hence, it equals $|F(z)| = |z^N - 1|$. Note what this gives in the $N \to \infty$ limit. If z equals any of the Nth roots of 1, we obtain zero, of course (for any N). But if z is any point inside the unit circle, its magnitude is less than 1, so $|z^N| = |z|^N \to 0$. The product of the lengths is therefore $|0 - 1| = 1$, independent of z (and N, provided that $N \to \infty$). If z is any point outside the unit circle, then $|z^N| \to \infty$, so we obtain ∞. If z is exactly *on* the unit circle (but not equal to an Nth root of 1), then $|z^N - 1|$ doesn't approach a unique value, because the complex number z^N simply runs around the unit circle as N increases. All we can say is that $|z^N - 1|$ (which is the distance from z^N to 1) takes on values ranging from 0 to 2. ♣

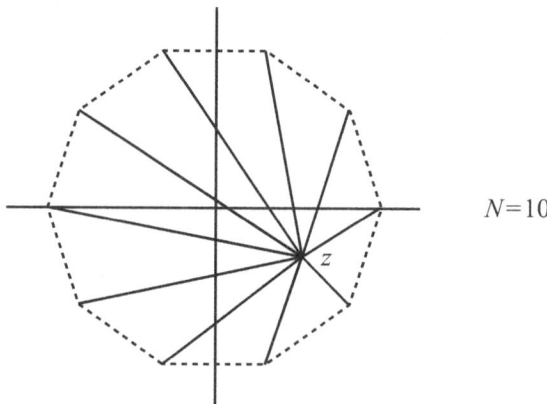

Figure 3.38

22. **Mountain climber**

CHEAP LASSO: We will take advantage of the fact that a cone is "flat," in the sense that we can make one out of a piece of paper without crumpling the paper. Cut the cone along a straight line emanating from the peak and passing through the knot of the lasso, and roll the cone flat onto a plane. The resulting object is a sector of a circle; call it S. See Fig. 3.39.

If the cone is very sharp, then S looks like a thin pie piece. If the cone is very wide, with a shallow slope, then S looks like a pie with a piece taken out of it. Points on the straight-line boundaries of the sector S are identified with each other.

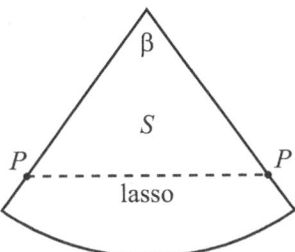

Figure 3.39

Let P be the location of the lasso's knot. Then P appears on each straight-line boundary, at equal distances from the tip of S. Let β be the angle of the sector S.

The key to this problem is to realize that the path of the lasso's loop must be a straight line on S, as shown by the dashed line in Fig. 3.39. (This is true because the rope takes the shortest path between two points since there is no friction, and rolling the cone onto a plane doesn't change distances.) But a straight line between the two identified points P is possible if and only if the sector S is smaller than a half-disk. The condition for a climbable mountain is therefore $\beta < 180°$.

What is this condition, in terms of the given angle α of the peak? Let C denote a horizontal cross-sectional circle of the mountain, a distance d (measured along the cone) from the top. (We are considering this circle for geometrical convenience. It is *not* the path of the lasso; see the remark below.) A half-disk sector S implies that when the cone is rolled onto a plane, the circle C becomes a semicircle, as shown in Fig. 3.40.

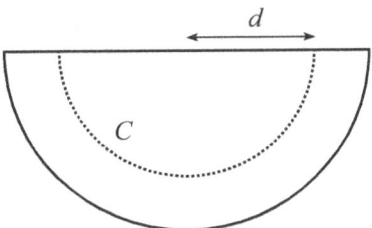

Figure 3.40

The circumference of C is therefore the length of the semicircle, which is πd. This then implies that the radius of the circle C on the cone is given by $2\pi r = \pi d \implies r = d/2$. Since S must be smaller than a half-disk, we see that the radius r of C must be smaller than $d/2$. Looking at the cone from the side tells us that $\sin(\alpha/2) = r/d$. Therefore, the condition that S is less than a half-disk is

$$\sin(\alpha/2) < \frac{d/2}{d} = \frac{1}{2} \implies \alpha < 60°. \tag{3.82}$$

This is the condition under which the mountain is climbable. In short, having $\alpha < 60°$ (and hence $\beta < 180°$) guarantees that there is a loop around the cone with a shorter length than the distance straight to the peak and back.

REMARK: When viewed from the side, the rope will appear perpendicular to the side of the mountain at the point opposite the lasso's knot. A common mistake is to draw the side view shown in Fig. 3.41(a) and then conclude that the condition for a climbable mountain is $\alpha < 90°$. This is incorrect because the loop does not lie in a plane. (Equivalently, the side view of the loop isn't a straight line, as drawn in Fig. 3.41(a).) Lying in a plane, after all, implies an elliptical loop. (The intersection of a plane and a cone is an ellipse, provided that the plane isn't titled too much.) And an elliptical loop implies that the loop passes horizontally straight through the location of the knot (perpendicular to the page in the figure). This horizontal piece of the loop will not be able to apply an upward tension force on the knot. However, there must certainly exist an upward force component on the knot; this is what in turn holds the climber up. We conclude that the loop cannot be elliptical, and hence cannot be planar; it must have a kink (an abrupt change in direction) where the knot is. The correct side view looks something like Fig. 3.41(b), with the rope taking a curved path. If we had instead posed the problem with a planar, triangular mountain, then the condition would in fact be $\alpha < 90°$; Fig. 3.41(a) would be correct. On such a mountain, the loop has the necessary kink at the location of the knot. The loop basically doubles back along itself on the other side of the planar mountain. ♣

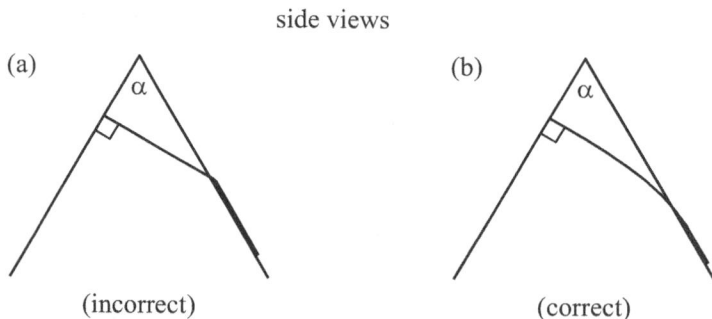

Figure 3.41

DELUXE LASSO: If the mountain is very steep, the climber will end up moving downward by means of the loop growing larger. (If the sector S in Fig. 3.39 is very thin, then sliding the loop down the mountain barely increases the loop's length; the dashed line between the P's barely gets any longer. The climber therefore falls essentially the same distance the knot falls.) If the mountain has a shallow enough slope, the climber will again end up moving downward, but now by means of the loop growing *smaller*. (If the sector S in Fig. 3.39 is nearly a half-disk, then sliding the loop *up* the mountain decreases the loop's length by essentially twice the distance the knot moves up toward the peak. See Fig. 3.42.

This decrease in the length of the loop gets added to the length of the rope hanging down to the climber. So if the knot moves up by ℓ, the climber moves down by 2ℓ relative to the knot. The net motion is therefore $2\ell - \ell = \ell$ downward along the mountain.) The only scenario in which the climber doesn't slide downward is the one where the change in position of the knot along the mountain is exactly compensated for by the change in length of the loop, so that the climber remains at the same height. We can find this special scenario as follows.

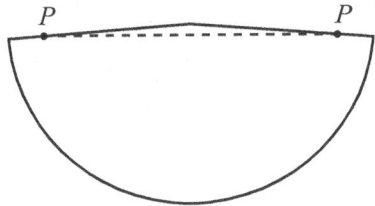

Figure 3.42

Roll the cone flat onto a plane, as we did in the cheap-lasso case. In terms of the sector S in the plane, the above condition requires that if we move the point P (the knot) a distance ℓ up (or down) along the mountain, the distance between the identified points P in Fig. 3.39 must decrease (or increase) by ℓ. If P moves a distance ℓ up the mountain, then the distance between the identified points P decreases by $2 \cdot \ell \sin(\beta/2)$. Setting this equal to ℓ gives $\sin(\beta/2) = 1/2 \Longrightarrow \beta = 60°$. The two P's and the tip of the sector S in Fig. 3.39 therefore form an equilateral triangle.

What peak-angle α does $\beta = 60°$ correspond to? As in part the cheap-lasso case, let C be a cross-sectional circle of the mountain, a distance d (measured along the cone) from the top. Then the $\beta = 60°$ modification of Fig. 3.40 tells us that the circumference of C equals $(\pi/3)d$. This implies that the radius of the circle C on the cone is given by $2\pi r = (\pi/3)d \Longrightarrow r = d/6$. Looking at the cone from the side then gives

$$\sin(\alpha/2) = \frac{d/6}{d} = \frac{1}{6} \quad \Longrightarrow \quad \alpha \approx 19°. \tag{3.83}$$

This is the condition under which the mountain is climbable. We see that there is exactly one angle for which the climber can climb up along the mountain, in contrast with the whole $\alpha < 60°$ range for the cheap lasso. The cheap lasso is therefore much more useful than the fancy deluxe lasso, assuming, of course, that you want to use it for climbing mountains, and not, say, for roping cattle.

Another way to derive the $\beta = 60°$ result for the deluxe lasso is to note that the three directions of rope emanating from the knot all have the same tension, because the deluxe lasso is one continuous piece of rope (with no friction at the knot). The three directions must therefore have 120° angles between themselves. (This is the only way that the three equal tensions can provide zero net force on the massless knot, as you can verify.) This then implies that $\beta = 60°$ in Fig. 3.39.

EXTENSION: For each type of lasso, we can also ask the question: For what angles can the mountain be climbed if the lasso is looped N times around the top of the mountain? The solution here is similar to the one for the $N = 1$ case in the original problem. You should set the book aside and try to solve this extension before reading further.

For the cheap lasso, roll the cone N times onto a plane, as shown in Fig. 3.43 for $N = 4$. The resulting shape S_N is a sector of a circle divided into N equal sectors, each representing a copy of the cone. The path of the lasso is still a straight line in the plane, so as in the original solution, S_N must be smaller than a half-disk. The circumference of the circle C (defined earlier) must therefore be less than $\pi d/N$. Hence, the radius of C must be less than $(\pi d/N)/2\pi = d/2N$. So instead of Eq. (3.82) we now have

$$\sin(\alpha/2) < \frac{d/2N}{d} = \frac{1}{2N} \implies \alpha < 2\sin^{-1}\left(\frac{1}{2N}\right). \quad (3.84)$$

For $N = 1$, this gives $\alpha < 60°$, as we found in the original solution.

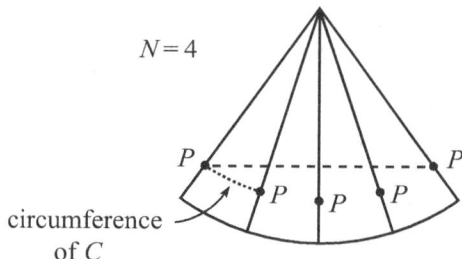

Figure 3.43

For the deluxe lasso, again roll the cone N times onto a plane. From the same reasoning as before, the $\beta = 60°$ result now becomes $N\beta = 60°$. The circumference of C must therefore be $(\pi/3)d/N$, which means that its radius must be $d/6N$. So instead of Eq. (3.83) we now have

$$\sin(\alpha/2) = \frac{d/6N}{d} = \frac{1}{6N} \implies \alpha = 2\sin^{-1}\left(\frac{1}{6N}\right). \quad (3.85)$$

For $N = 1$, this gives $\alpha \approx 19°$, as we found above.

23. **Passing the spaghetti**

   (a) Let's solve the problem for a few small values of n, to get a feel for things.

   If $n = 2$, the one person not at the head of the table is guaranteed to be the last served (LS).

   If $n = 3$, the two people not at the head have equal $1/2$ chances of LS.

   If $n = 4$, let's label the diners as A, B, C, D (with A being the head), going cyclically around the table. Consider D's probability of LS. The various paths of spaghetti that allow D to be the last served are:

$$\text{ABC}\ldots, \quad \text{ABABC}\ldots, \quad \text{ABABABC}\ldots, \quad \text{etc.} \tag{3.86}$$

At any stage, there is a 1/2 chance of passing the plate to a given neighbor. So the probability of the ABC path is $1/2^2$, the probability of ABABC is $1/2^4$, and so on, with an additional factor of $1/2^2$ tacked on for each additional AB pair in the sequence. The sum of the probabilities of all the above paths is therefore

$$\frac{1}{2^2} + \frac{1}{2^4} + \frac{1}{2^6} + \cdots = \frac{1/4}{1 - 1/4} = \frac{1}{3}. \tag{3.87}$$

By symmetry, B also has a 1/3 chance of LS, which then leaves a 1/3 chance for C. Hence, B, C, and D all have equal 1/3 chances of LS.

The probabilities for $n = 5$ are a bit tedious to calculate in the same manner (but doable if you feel inspired to work them out). So at this point we will (for lack of a better option) make the following claim:

Claim: *For arbitrary n, all diners not at the head of the table have equal $1/(n - 1)$ probabilities of being the last served (LS).*

This seems a bit counterintuitive, because you might think that the diners farther from the head have a greater chance of LS. But the claim is in fact correct.

Proof: A necessary and sufficient condition for a given diner to be the last served is that the following two things happen:

(1) The plate must approach the given diner from the right or left and reach the person next to that diner (or start there, if the diner is located right next to the head).

(2) The plate must then reverse its direction and make its way (in whatever manner, as long as it never touches the given diner) all the way around the table until it reaches the person on the other side of the given diner.

For any of the (non-head) diners, the probability that the first of these events will eventually happen is 1. This event therefore doesn't differentiate between the $n - 1$ (non-head) probabilities of LS.

Given that event #1 has happened, there is some definite probability of event #2 happening, independent of where the given diner is located. This is true because the probability of traveling all the way around the table (from the person on one side of the given diner to the person on the other) doesn't depend on where this traveling starts. (It is irrelevant who has already received some spaghetti. So all non-head diners are equivalent, as far as event #2 goes.) Hence, event #2 also doesn't differentiate between the $n - 1$ (non-head) probabilities of LS.

Therefore, since neither of the events produces any differentiation, all of the $n - 1$ (non-head) probabilities of LS are the same and are thus equal to $1/(n - 1)$. ∎

(b) This problem is equivalent to asking how many steps it takes, on average, for a random walk in one dimension to hit its nth site.

Let f_n be the expected number of steps. And let g_k be defined as follows. Assume that k sites have been visited, and that the present position is at one of the ends of the string of these k sites. Define g_k to be the expected number of steps it takes to reach a new site. We then claim that

$$f_n = f_{n-1} + g_{n-1}. \tag{3.88}$$

This is true because in order to reach n sites, you must first reach $n-1$ sites; this takes f_{n-1} steps, on average. (And you are now at an end of the string of $n-1$ sites.) You must then reach one more site, starting at an end of the string of $n-1$ sites; this takes g_{n-1} steps, on average.

Claim: $g_n = n$.

Proof: Let the sites that have been visited be labeled $1, 2, \ldots, n$. (These numbers represent the order along the line, not the order visited.) Let the present position be site 1.

There is a $1/2$ chance that the next step will be to site 0, in which case it takes only one step to reach a new site.

There is a $1/2$ chance that the next step will be to site 2. By considering this site to be an end-site of the string $2, 3, \ldots, n-1$ (which has length $n-2$), we see that it takes g_{n-2} steps (on average) to reach sites 1 or n. And then from each of these, it takes of g_n steps (on average) to reach a new site.

Putting the above two results together gives

$$g_n = \frac{1}{2} \cdot 1 + \frac{1}{2} \cdot (1 + g_{n-2} + g_n) \implies g_n = g_{n-2} + 2. \tag{3.89}$$

Since g_1 is simply 1, and since a slight tweak to the above reasoning gives $g_2 = (1/2)(1) + (1/2)(1 + g_2) \implies g_2 = 2$ (equivalently, $g_0 = 0$), we inductively obtain $g_n = n$. ∎

Using $g_n = n$, Eq. (3.88) becomes $f_n = f_{n-1} + (n-1)$. Starting with $f_1 = 0$ (or $f_2 = 1$), we see inductively that f_n is the sum of the first $n-1$ integers. Hence,

$$f_n = \frac{n(n-1)}{2}. \tag{3.90}$$

24. How many trains?

(a) The fact that the trains arrive randomly means that the occurrences on one track are completely independent of the occurrences on all the other tracks. So the probability that an approaching train is your train is $1/n$. Therefore, the probability that the first $k-1$ trains to arrive (starting at an arbitrary random moment in time) are not yours, while the kth one is, equals $(1 - 1/n)^{k-1}(1/n)$. The average number of trains that have come by the time yours arrives (including yours) is then

$$A = \sum_{k=1}^{\infty} k \left(1 - \frac{1}{n}\right)^{k-1} \left(\frac{1}{n}\right). \tag{3.91}$$

If we define $x \equiv 1 - 1/n$, we can write A as

$$A = (1 - x)(1 + 2x + 3x^2 + 4x^3 + \cdots). \tag{3.92}$$

There is a standard method for calculating the sum here. We can break it up into the sum of an infinite number of infinite geometric series:

$$1 + 2x + 3x^2 + 4x^3 + \cdots \tag{3.93}$$
$$= (1 + x + x^2 + x^3 + \cdots) + (x + x^2 + x^3 + \cdots) + (x^2 + x^3 + \cdots) + \cdots,$$

which itself equals an infinite geometric series (making use of the $a_0/(1-r)$ sum formula):

$$\frac{1}{1-x} + \frac{x}{1-x} + \frac{x^2}{1-x} + \cdots = \frac{1}{(1-x)^2}. \tag{3.94}$$

(Alternatively, this result follows from taking the derivative of the relation, $1/(1-x) = 1 + x + x^2 + x^3 + \cdots$.) We therefore have

$$A = (1-x)\frac{1}{(1-x)^2} = \frac{1}{1-x} = \frac{1}{1-(1-1/n)} = n. \tag{3.95}$$

The expected number of trains you see is therefore n.

(b) Our goal now is to calculate the average number of trains you see (by the time yours arrives), starting right after the most recent arrival of your train. But the reasoning in part (a) is independent of the starting time; it works for any starting time, including right after the arrival of one of your trains. So the answer is the same. That is, on average your train will be the nth one to arrive after the previous arrival of your train.

Here is another line of reasoning: Let a very large number N of trains come by. Let A be the desired average number of trains you see (by the time yours arrives), starting right after the most recent arrival of your train. If x of the N trains are yours, then we have $N \approx Ax$ (with the "$\approx$" becoming more of an "=" the larger N is). But if you see x of your trains, then a person waiting on any one of the other n tracks will also see (approximately) x of her trains. Therefore $N \approx nx$. Equating our two expressions for N gives $A = n$.

You might wonder how the answers to parts (a) and (b) can be the same, due to the following reasoning that suggests that the answer to part (b) should be twice the answer to part (a). If you arrive at the train station at a random time (as in part (a)), and if you have to wait an average of n trains for yours to arrive, then you must also have arrived an average of n trains after your previous train arrived. (Any reasoning that is valid for future occurrences is also valid for past occurrences. Imagine listing out a long string of trains; there is no difference between forward in time and backward in time.) So the total number of trains between two successive arrivals of your train should be $2n$, on average. Why doesn't this answer of $2n$ agree with the answer of n that we obtained above for part (b)? See Problem 45 for a discussion of this. (That one is a classic; don't peek at the answer too soon.)

25. **Flipping a coin**

(a) The various outcomes of the game are T, HT, HHT, HHHT, ... These occur with probabilities 1/2, 1/4, 1/8, 1/16, ..., and your respective winnings are 1, 2, 3, 4, ... dollars. The expected value of your winnings is therefore

$$\frac{1}{2} + \frac{2}{4} + \frac{3}{8} + \frac{4}{16} + \cdots. \tag{3.96}$$

This can be written as the sum of an infinite number of infinite geometric series:

$$\left(\frac{1}{2} + \frac{1}{4} + \frac{1}{8} + \frac{1}{16} + \cdots\right) + \left(\frac{1}{4} + \frac{1}{8} + \frac{1}{16} + \cdots\right) + \left(\frac{1}{8} + \frac{1}{16} + \cdots\right) + \cdots, \tag{3.97}$$

which itself equals an infinite geometric series (making use of the $a_0/(1-r)$ sum formula):

$$(1) + \left(\frac{1}{2}\right) + \left(\frac{1}{4}\right) + \cdots = 2. \tag{3.98}$$

So you can expect to win an average of two dollars each time you play this game.

(b) Your winnings in the T, HT, HHT, HHHT, ... cases are now 1, 2, 4, 8, ... dollars. So the expected value of your winnings is

$$\frac{1}{2} + \frac{2}{4} + \frac{4}{8} + \frac{8}{16} + \cdots = \frac{1}{2} + \frac{1}{2} + \frac{1}{2} + \frac{1}{2} + \cdots = \infty. \tag{3.99}$$

You will quickly discover, however, that you will not win an infinite amount of money playing this game. We therefore seem to have a paradox. The expectation value is infinite, but certainly no one is going to put up an infinite amount of money, or even a million dollars, for the opportunity to play the game once. What is the resolution to this paradox? (You should ponder this deeply before reading further.)

The resolution is that an expectation value is defined to be an average over an *infinite* number of trials (or the limit toward an infinite number). But you are simply not going to play an infinite number of games. In other words, the calculated expectation value doesn't agree with your experiment, because your experiment has nothing to do with the precise definition of an expectation value. To be sure, if you somehow did play an infinite number of games, then you would indeed have an infinite average for your winnings. The paradox arises from trying to make "expectation value" mean something it doesn't.

This might not be a very satisfying explanation, so let's get a better feel for the paradox by looking at a situation where someone plays $N = 2^n$ games. How much money would a "reasonable" person be willing to put up for the opportunity to play these N games?

Well, in about half of the games (2^{n-1} of them) the person will win one dollar; in about a quarter of them (2^{n-2}) she will win two dollars; in about an eighth of them (2^{n-3}) she will win four dollars; etc., until in about $1/2^n$ of them (one game) she will win 2^{n-1} dollars. In addition, there are the "fractional" numbers of games where she wins much larger quantities of money. For example, in about half a game (on average) she will win 2^n dollars, etc. These fractional games are where the infinite expectation value comes from, as we saw in Eq. (3.99). But let's forget about these for the moment, in order to just get a lower bound on what a reasonable person should put on the table (leaving aside complicated issues such as the marginal utility of money). Adding up the above non-fractional cases gives the winnings as $2^{n-1}(1) + 2^{n-2}(2) + 2^{n-3}(4) + \cdots + 1(2^{n-1}) = n2^{n-1}$. The average value of these winnings in the $N = 2^n$ games is therefore $n2^{n-1}/2^n = n/2$. Since $n = \log_2 N$, we can write this in terms of N as $(\log_2 N)/2$.

A reasonable person should therefore expect to win at least $(\log_2 N)/2$ dollars per game. (By "expect," we mean that if the player plays a very large number of sets of N games, and then takes an average over all the games in these sets, she will win at least $(\log_2 N)/2$ per game.) This increases with N and goes to infinity as N goes to infinity (although slowly, like a log). It is nice to see that we can obtain this infinite limit without having to worry about what happens in the infinite number of "fractional" games. Remember, though, that this quantity, $(\log_2 N)/2$, has nothing to do with a true expectation value, which is defined only for $N \to \infty$ and which in the present problem is infinite.

Someone might still not be satisfied with this explanation and want to ask, "But what if I play only N games? I will never ever play another game. How much money do I expect to win?" The proper answer is that the question has no meaning. It is not possible to define how much one *expects* to win, if one is not willing to take an average over a arbitrarily large number of trials.

26. **Trading envelopes**

(a) Let your envelope contain N dollars. Then the other envelope contains either $2N$ or $N/2$ dollars, with equal chances of each. If you switch, the expectation value of your assets is $\frac{1}{2}(2N) + \frac{1}{2}(N/2) = 5N/4$. This is greater than N, so you should switch. In short, you stand to gain N but risk losing only $N/2$.

(b) There are (at least) two possible modes of reasoning, yielding different results:

- It seems that we should be able to use the same reasoning as in part (a). If you have N dollars in your envelope, then the other one contains either $2N$ or $N/2$ dollars. Since you have a 50-50 chance of picking either envelope, the other envelope should have a 50-50 chance of containing $2N$ or $N/2$ dollars. If you switch, there is a $1/2$ chance you win N dollars, and a $1/2$ chance you lose $N/2$ dollars. Therefore, the expectation value of your gain is $N/4$ dollars. So you should switch.

- If the correct strategy is to switch (that is, if there is an average gain from trading), then if person A picks one envelope and person B picks the other, they are both better off if they switch. This cannot be true. Likewise, it cannot be true that they are both better off if they do not switch. Therefore, it doesn't matter whether or not they switch.

The second reasoning is correct. The flaw in the first reasoning is that the other envelope does *not* have a 50-50 chance of containing $2N$ or $N/2$ dollars. Such a 50-50 distribution would yield a zero probability of the envelopes containing a finite and nonzero quantity (as we'll explain below). In a nutshell, it is incorrect to assume that because you have a 50-50 chance of picking either envelope, the other envelope has a 50-50 chance of containing twice or half the amount in your envelope.

If you want to be explicit in the same manner as in part (a), then let the two envelopes contain N and $2N$ dollars. And assume that your strategy is to switch. There is a $1/2$ chance that you start with N, in which case you gain N if you switch. And there is a $1/2$ chance that you start with $2N$, in which case you lose N if you switch. So the expectation value of your gain is $\frac{1}{2}(N) + \frac{1}{2}(-N) = 0$. In short, you gain N half the time and lose N half the time (in contrast with the gain of N but loss of $N/2$ in part (a)).

(c) As we have stated, the fundamental difference between the scenarios in parts (a) and (b) is that the second envelope in scenario (b) does not have a 50-50 chance of containing twice or half the amount in your envelope. Let's see why.

Consider the following slightly modified game, which has all the essentials of the original one. Consider a game where powers of 2 (positive, negative, or zero) are the only numbers of dollars allowed in the envelopes.

To see why in scenario (b) there isn't a 50-50 chance that the other envelope contains $2N$ or $N/2$ dollars, let's look at the simplest distribution of money in the envelopes, the case where only two quantities are used. Let's say that I always put $4 in one envelope and $8 in the other. (And assume that you have a bad memory and can't remember anything from one game to the next.) If your strategy is to switch, and if you initially have $4, then you will definitely win $4 on the switch. And if you initially have $8, then you will definitely lose $4 on the switch. Since you have a 50-50 chance of starting with the $4 or $8 envelope, you will on average neither win nor lose any money. In this example, it is clear that if you have, for example, the $4 envelope, there is *not* a 50-50 chance that the other envelope contains $2 or $8. Instead, there is a 100% chance that it contains $8.

You can try to make a situation in scenario (b) that comes "close" to always having a 50-50 chance that the other envelope contains twice or half the amount in your envelope. For example, let there be a $1/n$ chance that the envelopes contain 2^k and 2^{k+1} dollars, for all k from 1 to n. Then indeed if there are 2^m dollars in your envelope, for $m = 2, \ldots, n-1$, there is a 50-50 chance that the other envelope contains twice or half that amount. In all of these $n-2$ cases, you will win money, on average, if you switch. And you will certainly win money if you switch in the case where you have the minimum amount, 2^1 dollars. You will, however, lose a great deal of money (2^n dollars, in fact) if you happen to start out with 2^{n+1} dollars. This happens only $1/(2n)$ of the time, but it in fact precisely cancels, on average, the winnings from all the other $n-1$ cases (as you can show). Therefore, it doesn't matter if you switch.

If you want to produce a 50-50 chance that the other envelope contains twice or half the amount in your envelope, for *all* values of m, then you have to assign equal probabilities to all of the $(2^k, 2^{k+1})$ pairs, for $-\infty < k < \infty$. But the assignment of equal probabilities to an infinite set requires that all of the probabilities are zero, which means that there is a zero chance of putting a finite amount of money in the envelopes. Since it is stated that there *is* some amount of money in the envelopes, we conclude that the probabilities of the $(2^k, 2^{k+1})$ pairs are not all equal. The setup in part (b) is therefore not the same as in part (a) (where the other envelope *does* have a 50-50 chance of containing twice or half the amount in your envelope), so there is no paradox.

27. Waiting for an ace

FIRST SOLUTION: In the general case where the total number of cards is N and the number of aces is n, the answer is $(N+1)/(n+1)$. (So the answer to the stated problem is $53/5 = 10.6$.) This general result can be conjectured by playing around with some small values of N and n (as you should do). We can prove it by induction on N, as follows.

Let's add a non-ace card to the deck. So there are now $N+1$ cards, n of which are aces. If we start dealing cards, there are two possibilities: (1) There is an $n/(N+1)$ chance that the first card is an ace, in which case we have to deal only one card. (2) There is a $1 - n/(N+1)$ chance that the first card isn't an ace, in which case we now have a deck of the original type in our hand (N cards, n of which are aces). From the inductive hypothesis, we then have to deal an average of $(N+1)/(n+1)$ cards to get an ace. Adding on the initial card we dealt, we see that we need to deal a total average number of $1 + (N+1)/(n+1)$ cards to get an ace in this second case.

Combining the above two cases, the expected number of cards needed to get an ace is

$$\frac{n}{N+1} \cdot 1 + \left(1 - \frac{n}{N+1}\right) \cdot \left(1 + \frac{N+1}{n+1}\right). \tag{3.100}$$

You can show that this simplifies to $(N+2)/(n+1)$, which takes the form of the conjecture, with N replaced by $N+1$. This completes the induction on N.

In the case where $N = n$ (that is, all the cards are aces), the $(N+1)/(n+1)$ expression for the expected number of cards equals 1. This is correct, of course, because the first card is guaranteed to be an ace. Now imagine picking an arbitrary value of n. Since the $(N+1)/(n+1)$ conjecture is correct for $N = n$ (which is the smallest possible value of N, given n), and since it is also correct for all larger N by the inductive step, it is therefore coorect for all N ($\geq n$), for any n.

SECOND SOLUTION: Add an $(n + 1)$th ace to the given deck. There are now $N + 1$ cards in all, $n + 1$ of which are aces. Randomly place the $N + 1$ cards in a circle. Let the location of the $(n + 1)$th ace be the starting point. The location of this card is random, so this is as random a starting point as any. (By "starting point," we mean that the next card in the clockwise direction is the first card you deal. The card after that one is the second card you deal, etc.) The average clockwise distance to the next ace (which is the answer we're looking for) is the same as the average distance between any other successive aces. There are $n + 1$ ace-to-ace intervals in all, and there are $N + 1$ cards, so the common average distance from one ace to the next is $(N + 1)/(n + 1)$. This is the average number of cards you need to deal to get your first ace.

REMARK: If you're completely comfortable with the above strategy of adding an $(n + 1)$th ace to the given deck, then feel free to skip this remark. But if the strategy seems a little sneaky, then read on.

Why can't we just put the original N cards (n of which are aces) in a circle? The average distance between the aces is then simply N/n. However, although this statement is true, it doesn't help us solve the problem. You might think that the answer should just be N/n, because you can pick any one of the aces to be the starting point. But that is incorrect, because the setup then corresponds to a deck with $n - 1$ aces and $N - 1$ cards total; the starting card doesn't count as part of the deck. Alternatively, you might think that the answer should be $N/2n$, because if you pick a random starting point, it should be (on average) in the middle of an interval with an average length of N/n. This is also incorrect, for the reasons discussed in Problem 45. (That one is a classic; don't peek at the answer too soon.)

To make the strategy of adding an $(n + 1)$th ace more palatable, consider the following setup, which is essentially equivalent to our setup with the aces. Imagine randomly throwing n darts onto a segment with length ℓ, with uniform probability of hitting anywhere along the segment. These darts will divide the segment into $n + 1$ pieces. We claim that if you repeat this process many times, all of the pieces will have the same average length of $\ell/(n + 1)$. (For example, the 5th interval from the left end will have an average length of $\ell/(n + 1)$. Likewise for the 8th, etc.) On one hand, you might think this is obvious. On the other hand, you might think that the interior pieces (between two darts) will each have an average length of ℓ/n, while the outer two pieces (between a dart and an end) will each have an average length of $\ell/2n$. This latter view is incorrect, for the following reason.

Wrap the given segment into a circle. Throw a blue dart randomly at it, and then throw n red darts. By symmetry, the total of $n + 1$ darts divide the circle into $n + 1$ pieces with the same average length (the color of a dart doesn't matter here, of course), which must then be $\ell/(n + 1)$. (For example, the 5th piece, measured clockwise from the blue dart, will have an average length of $\ell/(n + 1)$, as will all the others.) Now cut the circle at the blue dart (and throw away that dart), and unwrap the circle into a line. This produces the linear-segment setup in the preceding paragraph. The pieces in the linear setup therefore all have the same average length of $\ell/(n + 1)$, even the two pieces at the ends.

If you're not comfortable having a random throw of the blue dart determine where the ends of the line segment are, you can start with the given line segment, then wrap it into a circle, and then specifically place the blue dart at the location where the ends join. This circular setup is certainly identical to the linear-segment setup. Of course, this placement of the blue dart isn't a random action, but it might as well be, because the blue dart has to end up *somewhere*, and all points on a circle are equivalent (at least before anything else has been thrown at it). Whether the blue dart is placed purposefully or thrown randomly, you can always just rotate the circle so that the blue dart is at the top, before you start throwing the red darts. The red darts therefore can't tell the difference between purposeful and random placement of the blue dart.

In the above discussion, we threw the blue dart first. But this isn't necessary; we can throw it anytime. In particular, we can throw it last. In this case, before we throw the blue dart, the n red darts break up the circle into pieces that all have the same average length of ℓ/n. (By this we mean that if we single out one of the red darts by putting a piece of tape on it, then, for example, the 5th piece, measured clockwise from the taped red dart, will have an average length of ℓ/n, as will all the others.) But then when we finally throw the additional blue dart, the pieces now all have the same average length of $\ell/(n+1)$, even though the blue dart split only one of the original pieces in two. If you find this bizarre, see Problem 45. ♣

THIRD SOLUTION: This is the brute-force solution. It isn't terribly enlightening, but we'll present it just in case you tried to solve the problem this way and were wondering if it's possible to evaluate the sum that arises. (Answer: Yes, but it's tricky.)

Consider the general case where the total number of cards is N and the number of aces is n. We can enumerate the various possibilities for when you get the first ace:

- There is an n/N chance that the first card is an ace.
- There is an $(N-n)/N$ chance that the first card isn't an ace, and then an $n/(N-1)$ chance that the second card is.
- There is an $(N-n)/N$ chance that the first card isn't an ace, then an $(N-n-1)/(N-1)$ chance that the second card also isn't, and then an $n/(N-2)$ chance that the third card is. And so on.

The products of the various factors in each of these cases gives the probability $p(k)$ that the kth card is your first ace. The expected number of cards you need to deal is then $C = \sum k \cdot p(k)$. By writing out the first few terms in the following sum, you can verify that C can be written as (the index k is the number of non-ace cards you deal before the ace)

$$C = \sum_{k=0}^{N-n} \frac{\binom{N-n}{k}}{\binom{N}{k}} \cdot \frac{n}{N-k} \cdot (k+1). \tag{3.101}$$

The goal now is to calculate this sum in closed form. If you throw it into Mathematica, you do indeed get the desired result of $(N+1)/(n+1)$. To proceed analytically, one method is the following. (This is how I did it; there might be a quicker way.) We start by noting the binomial identity,

$$\binom{N}{m}\binom{m}{k} = \binom{N}{k}\binom{N-k}{m-k}. \qquad (3.102)$$

To prove this, you can expand the binomial coefficients in terms of factorials. Or a better way is the following. The lefthand side is the number of ways to choose a committee of m people from a group of N people, and to then designate k of them as officers. But the righthand side represents the same end result, by first picking the k officers from the whole group of N people, and then picking the $m-k$ ordinary committee members from the remaining $N-k$ people.

With $m \equiv N - n$, Eq. (3.102) turns Eq. (3.101) into

$$C = \sum_{k=0}^{N-n} \frac{\binom{N-k}{N-n-k}}{\binom{N}{N-n}} \cdot \frac{n}{N-k} \cdot (k+1), \qquad (3.103)$$

which will be easier to deal with because the binomial coefficient in the denominator now doesn't involve the summation index k. You can show that Eq. (3.103) simplifies to (with $j \equiv k+1$)

$$C = \frac{1}{\binom{N}{N-n}} \sum_{k=0}^{N-n} \binom{N-1-k}{n-1} \cdot (k+1) = \frac{1}{\binom{N}{N-n}} \sum_{j=1}^{N-n+1} \binom{N-j}{n-1} \cdot j. \qquad (3.104)$$

Written out, the sum here is

$$\binom{N-1}{n-1} + 2\binom{N-2}{n-1} + 3\binom{N-3}{n-1} + \cdots + (N-n+1)\binom{n-1}{n-1}. \qquad (3.105)$$

We now need one more binomial identity:

$$\binom{a}{a} + \binom{a+1}{a} + \binom{a+2}{a} + \cdots + \binom{b}{a} = \binom{b+1}{a+1}, \qquad (3.106)$$

which you can prove by induction on b. (You should pause and do that now.) This differs from Eq. (3.105) in that Eq. (3.105) has the extra j factors in front of the binomial coefficients. But we can write Eq. (3.105) as the sum of Eq. (3.106)-type sums, each with a different number of terms; this will generate the appropriate j factors:

$$\binom{N-1}{n-1} + \binom{N-2}{n-1} + \binom{N-3}{n-1} + \cdots + \binom{n-1}{n-1}$$
$$+ \binom{N-2}{n-1} + \binom{N-3}{n-1} + \cdots + \binom{n-1}{n-1} \qquad (3.107)$$
$$+ \binom{N-3}{n-1} + \cdots + \binom{n-1}{n-1}$$
$$\vdots$$

In the notation of Eq. (3.106) with $a \equiv n - 1$, the first line of Eq. (3.107) is a sum up to $b = N - 1$ (which gives $\binom{N}{n}$ from Eq. (3.106)), the second line is a sum up to $b = N - 2$ (which gives $\binom{N-1}{n}$), the third line is a sum up to $b = N - 3$ (which gives $\binom{N-2}{n}$), and so on, down to a sum up to $b = n - 1$ (which consists of just one term, which can be written as $\binom{n}{n}$ to make it look similar to the other sums). Adding up all of these sums by again using Eq. (3.106) yields $\binom{N+1}{n+1}$. Eq. (3.104) then finally gives

$$C = \frac{\binom{N+1}{n+1}}{\binom{N}{N-n}} = \frac{\frac{(N+1)!}{(n+1)!(N-n)!}}{\frac{N!}{(N-n)!n!}} = \frac{N+1}{n+1}, \qquad (3.108)$$

as desired.

28. **Drunken walk**

(a) FIRST SOLUTION: Let the river and police station be located at positions 0 and N, respectively. Let $P_p(k)$ be the probability of ending up at the police station, given a present position of k. After one step, the drunk is equally likely to be at $k - 1$ or $k + 1$, so we have

$$P_p(k) = \frac{1}{2}P_p(k-1) + \frac{1}{2}P_p(k+1). \qquad (3.109)$$

Therefore, since the value of P at a given point equals the average of the values at the two neighboring points, $P_p(k)$ is a linear function of k. (Equivalently, Eq. (3.109) gives $P_p(k) - P_p(k-1) = P_p(k+1) - P_p(k)$, which means that P_p increases at a constant rate, from one k to the next.) Invoking the requirements that $P_p(0) = 0$ and $P_p(N) = 1$, we find that $P_p(n) = n/N$. The probability $P_r(n)$ of ending up at the river is then $P_r(n) = 1 - P_p(n) = 1 - n/N$. This can be written as $(N-n)/N$, so we see that the probability of ending up at a given end is proportional to the starting distance from the other end.

SECOND SOLUTION: Imagine a large number of copies of the given setup proceeding simultaneously. After each drunk takes his first step in all of the copies, the average position of all of them remains the same (namely, n steps from the river), because each one has a 50-50 chance of moving either way. The average position likewise remains unchanged after each successive step. This is true because the drunks who are still moving don't change the average position (because of their 50-50 random motion), and the drunks who have stopped at an end of the street certainly don't change the average position either (because they aren't moving). Therefore, the average position is always n steps from the river.

Let the drunks keep moving until all of them have stopped at either end. Let $P_r(n)$ and $P_p(n)$ be the probabilities of ending up at the river and police station, respectively, having started n steps from the river. Then after all the drunks have stopped, their average distance from the river is $0 \cdot P_r(n) + N \cdot P_p(n)$. But this must equal n. Hence, $P_p(n) = n/N$, and so $P_r(N) = 1 - (n/N)$.

(b) Let $g(k)$ be the expected number of steps it takes to reach an end (either one) of the street, given a present position of k. After one step, the drunk is equally likely to be at $k - 1$ or $k + 1$, so we have

$$g(k) = \frac{1}{2}g(k-1) + \frac{1}{2}g(k+1) + 1, \qquad (3.110)$$

where $g(0) = g(N) = 0$. We must now solve this recursion relation. Multiplying Eq. (3.110) through by 2, and then summing the resulting equation over values of k from 1 to m gives (as you can verify)

$$g(1) + g(m) = g(m+1) + 2m$$
$$\implies g(m+1) = g(1) + g(m) - 2m. \qquad (3.111)$$

Summing this relation over values of m from 1 to $n - 1$ gives

$$g(n) = n \cdot g(1) - 2\sum_{1}^{n-1} m$$
$$= n \cdot g(1) - n(n-1). \qquad (3.112)$$

So to find $g(n)$, we just need to determine $g(1)$. We can do this by setting $m = N - 1$ in Eq. (3.111), which yields $0 = g(1) + g(N-1) - 2(N-1)$. And since $g(1) = g(N-1)$ by symmetry, we find $g(1) = g(N-1) = N - 1$. Using this in Eq. (3.112) gives

$$g(n) = n(N-1) - n(n-1)$$
$$= n(N-n). \qquad (3.113)$$

This can be written as

$$g(n) = \left(\frac{N}{2}\right)^2 - \left(\frac{N}{2} - n\right)^2, \qquad (3.114)$$

which is just an inverted parabola. The maximum occurs at $n = N/2$ if N is even, or at $(N \pm 1)/2$ if N is odd (since n must be an integer).

REMARK: If you want to be a little more systematic about solving the recursion relation in Eq. (3.110), you can define $f(k) \equiv g(k) - g(k-1)$. (So f is the discrete derivative of g.) Multiplying Eq. (3.110) through by 2 and rearranging then gives $f(k+1) = f(k) - 2$. We see that the "derivative" f decreases linearly, which tells us that g is an inverted parabola (as we found above). As an exercise, you can solve for $f(k)$, and then $g(k)$, by using a strategy similar to the one we used above. ♣

29. HTH and HTT

(a) The HTH and HTT sequences both start with HT. Imagine a large number of different strings of coin flips. In each string, circle the first appearance of HT. The next letter is either an H or a T, with equal probabilities of $1/2$ for each. So we are guaranteed to get the first appearance of either HTH or HTT on the flip following the first HT pair. And then the game is over (because we are concerned here only with which sequence appears first). Therefore, since H and T occur with equal $1/2$ probabilities, the HTH and HTT sequences have equal $1/2$ probabilities of occurring first.

(b) It turns out, somewhat surprisingly, that even though the HTH and HTT sequences are equally likely to appear first, the average waiting times for the first occurrence of each sequence are *not* equal. This can be seen in the following way. As we saw above, in order to get either of the sequences, we must be at a location in the string where an HT appears. Consider the first appearance of HT. There are two equally likely possibilities for the next flip:

- *H appears next:* In this case we have obtained an HTH sequence, and we now need to wait for the first appearance of HTT. But note that we're already part of the way there, because the H that we just obtained might very well end up being the start of an HTT sequence. (In part (a), anything that happened after this point was irrelevant, because the game was over once we obtained the first sequence of either HTH or HTT.)
- *T appears next:* In this case we have obtained an HTT sequence, and we now need to wait for the first appearance of HTH. But since we just obtained a T, this doesn't help us in obtaining an HTH sequence.

We therefore see that although HTH and HTT are equally likely to occur first, we need to wait a shorter amount of time for an HTT sequence to appear after HTH occurs first (which happens 50% of the time), compared with the amount of time for an HTH sequence to appear after HTT occurs first (which also happens 50% of the time). We therefore have

$$E_{\text{HTH}} > E_{\text{HTT}}. \tag{3.115}$$

We have used the fact that the waiting time for HTH *if* it appears first is the same as the waiting time for HTT *if* it appears first. (This follows from the fact that each of these waiting times is one flip longer than the waiting time for the first HT pair.) So any difference between E_{HTH} and E_{HTT} is due only to the waiting time *after* the first occurrence of the other triplet.

(c) Let E_{HT} be the average waiting time for the first HT pair to occur. We'll calculate E_{HT} below, but for now let's see how E_{HTH} and E_{HTT} are related to E_{HT}.

Let's look at E_{HTH} first. After getting the first HT, there are two equally likely possibilities for the next flip:

- *H appears next:* In this case we have obtained the desired HTH sequence, so the expected total number of flips is $E_{\text{HT}} + 1$.

- *T appears next:* This T doesn't help us in eventually getting an HTH sequence, so we need to start the whole process over on the next flip. The T flip is wasted, and once we start over after the T flip, the expected waiting time from that point onward is E_{HTH}, by definition. So the expected total number of flips is $E_{\text{HT}} + 1 + E_{\text{HTH}}$.

Putting these two possibilities together (each of which occurs with probability 1/2) yields

$$E_{\text{HTH}} = \frac{1}{2}\left(E_{\text{HT}}+1\right) + \frac{1}{2}\left(E_{\text{HT}}+E_{\text{HTH}}+1\right) \implies E_{\text{HTH}} = 2E_{\text{HT}}+2. \quad (3.116)$$

Now let's look at E_{HTT}. After getting the first HT, there are two equally likely possibilities for the next flip:

- *T appears next:* In this case we have obtained the desired HTT sequence, so the expected total number of flips is $E_{\text{HT}} + 1$.

- *H appears next:* This H helps us in our goal of obtaining an HTT, because it might be the start of an HTT sequence. We've started a new string of letters, but with the advantage of a known H at the beginning. So we don't need to wait for as long as E_{HTT} after this point. How much shorter do we have to wait? In other words, how much benefit do we get from starting a string with a known H compared with starting with an unknown letter? It turns out that in the unknown-letter case, the first H appears on average on the second flip (we'll show this below). So we've saved ourselves one flip by having a known H on the first flip. Therefore, the expected waiting time (to get an HTT) after the first HT pair (in the event that the next flip is an H) is $E_{\text{HTT}} - 1$. The expected total number of flips is then $E_{\text{HT}} + (E_{\text{HTT}} - 1)$.

We can also write this result as $(E_{\text{HT}} + 1) + (E_{\text{HTT}} - 2)$. The first of the two terms here is the waiting time to get the H following the HT (assuming that we do get this H). And the second term is the additional waiting time to get an HTT, given that we already have an H (since it would take two flips, on average, to get that H).

The above two ways of writing the result (for the expected total number of flips, given that H appears after the first HT) differ in that $E_{\text{HTT}} - 1$ is the additional waiting time *including* the H (that is, starting right after the HT), and $E_{\text{HTT}} - 2$ is the additional waiting time *after* the H (that is, starting after the H flip after the HT).

REMARK: Concerning the above $E_{\text{HTT}} - 2$ result, it is indeed valid to simply subtract off 2 from E_{HTT}, when finding the waiting time after the H. To see why, imagine playing a million games and each time writing down the number of flips it takes to get the first HTT. You will generate a long list of numbers, the average of which is E_{HTT}. In each game, the average number of flips it takes to get the first H is $E_{\text{H}} = 2$. So the average number of flips between the first H and the first HTT must be $E_{\text{HTT}} - 2$. ♣

Putting these two possibilities together (each of which occurs with probability 1/2) yields

$$E_{\text{HTT}} = \frac{1}{2}(E_{\text{HT}} + 1) + \frac{1}{2}(E_{\text{HT}} + E_{\text{HTT}} - 1) \implies E_{\text{HTT}} = 2E_{\text{HT}}. \quad (3.117)$$

Eqs. (3.116) and (3.117) then quickly give

$$E_{\text{HTH}} = E_{\text{HTT}} + 2, \quad (3.118)$$

which is a nice result in itself.

Our task now reduces to finding the value of E_{HT}, that is, finding the average waiting time to obtain an HT pair. This can be done as follows. As we mentioned above (and as we'll show below), the average waiting time to obtain an H is 2 flips. After getting the first H, there are two equally likely possibilities for the next flip:

- *T appears next:* In this case we have obtained the desired HT pair, so the expected total number of flips is $2 + 1$.
- *H appears next:* In this case we need to start the process over, but with the advantage of having an H at the beginning, which might end up being the start of the desired HT pair. As above, this advantage allows us to subtract one flip from the waiting time (post-first-H; that is, including the new H), which would have been E_{HT} otherwise. So the expected total number of flips is $2 + (E_{\text{HT}} - 1)$.

Putting these two possibilities together (each of which occurs with probability 1/2) yields

$$E_{\text{HT}} = \frac{1}{2}(2 + 1) + \frac{1}{2}(2 + E_{\text{HT}} - 1) \implies E_{\text{HT}} = 4. \quad (3.119)$$

Eqs. (3.116) and (3.117) then give

$$E_{\text{HTH}} = 10 \quad \text{and} \quad E_{\text{HTT}} = 8. \quad (3.120)$$

REMARK: Let's now show that the average waiting time, E_{H}, for a single H is 2 flips. This can be done in (at least) three ways.

FIRST METHOD: If an H occurs for the first time on the nth flip, then the first $n-1$ flips must be Tails. A waiting time of four flips, for example, arises from the sequence TTTH, which occurs with probability $(1/2)^4 = 1/16$. The expectation value of the number of flips equals the sum of the products of the probabilities and the associated waiting times. So we have

$$E_H = \frac{1}{2} \cdot 1 + \frac{1}{4} \cdot 2 + \frac{1}{8} \cdot 3 + \frac{1}{16} \cdot 4 + \frac{1}{32} \cdot 5 + \cdots. \tag{3.121}$$

We can write this as the sum of an infinite number of infinite geometric series:

$$\begin{aligned} E_H = \frac{1}{2} &+ \frac{1}{4} + \frac{1}{8} + \frac{1}{16} + \frac{1}{32} + \cdots \\ &+ \frac{1}{4} + \frac{1}{8} + \frac{1}{16} + \frac{1}{32} + \cdots \\ &\quad\quad + \frac{1}{8} + \frac{1}{16} + \frac{1}{32} + \cdots \\ &\quad\quad\quad\quad + \frac{1}{16} + \frac{1}{32} + \cdots \\ &\quad\quad\quad\quad\quad\quad \vdots \end{aligned} \tag{3.122}$$

This has the correct number of each type of term. For example, the "1/16" appears four times. The first line is an infinite geometric series that sums to $a_0/(1-r) = (1/2)/(1-1/2) = 1$. The second line is also an infinite geometric series, and it sums to $(1/4)/(1-1/2) = 1/2$. Likewise the third line sums to $(1/8)/(1-1/2) = 1/4$. And so on. The sum of the infinite number of lines in the above equation therefore equals

$$E_H = 1 + \frac{1}{2} + \frac{1}{4} + \frac{1}{8} + \frac{1}{16} + \cdots. \tag{3.123}$$

But this itself is an infinite geometric series, and it sums to $a_0/(1-r) = 1/(1-1/2) = 2$, as we wanted to show.

SECOND METHOD: We can use the recursion type of argument that we used many times above (see Eqs. (3.116), (3.117), and (3.119)). There is a 1/2 chance that the first flip is an H, in which case it takes only one flip to get the H. There is a 1/2 chance that the first flip is a T, in which case we have to start the process over. We expect E_H flips after this point, and since we've already done one flip, the expected total waiting time is $1 + E_H$. Putting these two possibilities together (each of which occurs with probability 1/2) yields

$$E_H = \frac{1}{2}(1) + \frac{1}{2}(E_H + 1) \implies E_H = 2. \tag{3.124}$$

THIRD METHOD: Consider a long string of random H's and T's. Write down the number of flips it takes to get the first H. Then write down the number of flips it takes after that point to get the second H. Then write down the number of flips it takes after that point to get the third H. And so on. Each of these numbers is, on average, equal to E_H, by definition. So if we count out to the nth H in the string, on average this takes $n \cdot E_H$ flips. But we also know that on average it takes $2n$ flips, because on average half of the flips are H's. Therefore E_H must equal 2. ♣

(d) We found in Eq. (3.120) that the expected waiting times for the *first* appearances of HTH and HTT are 10 and 8, respectively. You might think that this implies that in a large number of flips, N, you should expect about $N/10$ appearances of HTH and $N/8$ appearances of HTT (remember that 10 and 8 are the expected waiting times until the *completion* of the sequence). So it seems like HTH should appear only 4/5 as often as HTT.

However, this is not correct. The error in the reasoning is that 10 and 8 are the expected waiting times for the *first* appearance of each sequence, but not necessarily the expected waiting times *between* successive appearances of each sequence. It turns out that although the expected waiting time between successive HTT's at any point in a string is in fact always 8 (because we have to start the game over after the second T), the expected waiting time between successive HTH's (more precisely, successive *completions* of HTH's) is 10 *only* for the first appearance. It then equals 8 for all subsequent appearances, as we will show. Since the difference in the expected waiting times for the first appearance is insignificant over the course of a large number N of flips, both sequences appear on average $N/8$ times.

Let us now show that once the first HTH has appeared, the expected waiting time between successive completions of HTH's is 8. The reasoning here is basically the same as the reasoning leading up to Eq. (3.117). Assume that we have just completed an HTH sequence (which means that our most recent flip is an H). The question we want to answer is: How long do we have to wait for the (completion of the) next HTH sequence, *given* that we've started with an H? As we saw above, the expected waiting time to get an H is 2. Therefore, since we have a known H at the moment in question (having just completed an HTH), and since we would normally have to wait 2 flips to obtain an H (if we had started from scratch), we see that the expected waiting time for the next HTH is 2 smaller than the $E_{HTH} = 10$ value we found in Eq. (3.120). So it equals $10 - 2 = 8$, as desired. (It is indeed valid to simply subtract off 2 from E_{HTH} here. The reasoning is the same as that given in the remark preceding Eq. (3.117), with HTT replaced with HTH.)

REMARKS:

1. As exercises, you can show that the expected waiting times for the *first* appearances of HHH and HHT are $E_{HHH} = 14$ and $E_{HHT} = 8$. And you can also show that the waiting times *between* successive appearances of these sequences are both 8. That is, both sequences appear, on average, $N/8$ times in a long string of N letters. The summary of our various results for the waiting times for the *first* appearance, and the waiting times *between* appearances, is therefore:

	HHH	HHT	HTH	HTT
first	14	8	10	8
between	8	8	8	8

The other four possible triplets (TTT, TTH, THT, THH) are equivalent to the four in the table; just switch the H and T labels.

Is there a good reason why the "between" waiting times for all eight triplets have the same value of 8? Indeed there is. Intuitively, for any given sequence of three letters, say HTH, there is a $(1/2)^3 = 1/8$ chance that a randomly chosen triplet in a string equals HTH. So the expected number of HTH's in a long string of N letters is $N/8$; hence the waiting time is 8. However, this reasoning is a little sloppy, because adjacent triplets aren't independent. For example, if the 36th, 37th, and 38th letters in a string form an HTH, then there is zero probability that the 37th, 38th, and 39th letters also form an HTH.

There are (at least) two ways to clean up the reasoning. We'll just sketch these; you can fill in the gaps. One way is to calculate the conditional probabilities that a given triplet (say, the 37th, 38th, and 39th letters) is an HTH, based on the possibilities for what the preceding triplet (the 36th, 37th, and 38th letters) is. You will obtain an overall probability of 1/8, assuming the earlier triplet is random. So you can start at the beginning and proceed inductively. Another way is to imagine listing out (in horizontal lines above each other) a very large number n of strings of N letters. Consider, say, the 37th, 38th, and 39th letters in each string. These triplets lie in a vertical column in the array. On average, $n/8$ of these triplets are HTH's. Likewise for any other column of triplets. (Imagine different people looking at the different columns, so that you don't need to worry about conditional probabilities.)

The above reasoning applies generally, not just to triplets. For example, if we instead consider quadruplets (like HTHH), then all of the $2^4 = 16$ possible quadruplets appear $N/16$ times, on average, in a long string of N letters. So the average waiting time between successive (completions of) appearances of any given quadruplet is $2^4 = 16$.

2. It is possible to use our recursion technique (applied many times throughout this problem) to calculate the expected waiting time for the first appearance of any sequence of any length. Consider, for example, the sequence HTHHH. We claim that

$$E_{\text{HTHHH}} = \frac{1}{2}\Big(E_{\text{HTHH}} + 1\Big) + \frac{1}{2}\Big((E_{\text{HTHH}} + 1) + (E_{\text{HTHHH}} - E_{\text{HT}})\Big).$$
(3.125)

(You should think about why this is true, before reading further.) The logic is: After getting the first HTHH, there is a 1/2 chance that the next flip is an H, in which case it takes only one additional flip to get the desired HTHHH; this is the first term in Eq. (3.125). The second term comes from the 1/2 chance that the next flip is a T, in which case we failed to get the desired HTHHH. However, although we failed, we have a possible head start in getting an HTHHH, because our most recent two letters, namely HT, are the beginning of a possible HTHHH sequence. So instead of having to start all over and wait an additional E_{HTHHH}, we only need to wait an additional $E_{HTHHH} - E_{HT}$, beyond the $E_{HTHH} + 1$ flips we've already done. (Yes, it is legal to simply subtract off E_{HT} from E_{HTHHH}, by reasoning similar to that given in the remark preceding Eq. (3.117).) Hence the second term in Eq. (3.125). Solving Eq. (3.125) for E_{HTHHH} gives

$$E_{HTHHH} = 2E_{HTHH} + 2 - E_{HT}. \qquad (3.126)$$

In general, the HT here is replaced by whatever letters are common to the end of the "failed" sequence and the start of the desired sequence (which is the same as the start of the failed sequence, of course). The failed sequence is the one obtained by switching the last letter in the desired sequence (H to T in the above example). So let's relabel E_{HT} as E_{fail}^{com} (for "common in failed sequence"). And let's relabel E_{HTHHH} with a general $E_{sequence}$, and E_{HTHH} with $E_{rem\ last}$ (for "remove last letter"). Then the more general version of Eq. (3.126) is

$$E_{sequence} = 2E_{rem\ last} + 2 - E_{fail}^{com}. \qquad (3.127)$$

As an example of a longer common sequence, the failed sequence for HTHHTHHH is HTHHTHHT, which has a common beginning/ending sequence of HTHHT. (It's fine if the beginning and ending sequences overlap.)

Eqs. (3.116) and (3.117) are special cases of the recursion relation in Eq. (3.127). In the former case, there are no common letters in the failed sequence. In the latter case, there is a common H, and $E_H = 2$. As an application of Eq. (3.127), we can start with $E_{HTH} = 10$ and work our way up to E_{HTHHH}. We have

$$E_{HTHH} = 2E_{HTH} + 2 - E_{HT} \implies E_{HTHH} = 2 \cdot 10 + 2 - 4 = 18. \qquad (3.128)$$

And then

$$E_{HTHHH} = 2E_{HTHH} + 2 - E_{HT} \implies E_{HTHHH} = 2 \cdot 18 + 2 - 4 = 34. \qquad (3.129)$$

3. Here is another way (and a more direct one, since it doesn't involve a recursion relation) to calculate $E_{sequence}$ for any sequence. We can use the fact (as we saw in the first remark above) that the expected waiting time *between* successive (completions of) appearances of any sequence is 2^n, where n is the length of the sequence. And as we saw in the solution to part (d), this 2^n waiting time between successive appearances may be shorter than the expected waiting time for the *first* appearance. This is due to the fact that upon completion of one sequence, we might have a head start in our quest for the next sequence, because the ending letter(s) of the sequence might be the same as the

beginning letter(s). (In part (d) with HTH, there was only a single common letter H.) Our waiting time between sequences (which we know is 2^n) is therefore shortened (relative to the waiting time E_{sequence} for the first appearance of the sequence) by the waiting time E_{com} for the common letters at the beginning and end.[11] That is

$$2^n = E_{\text{sequence}} - E_{\text{com}} \implies E_{\text{sequence}} = 2^n + E_{\text{com}}. \tag{3.130}$$

Note here that E_{com} deals with the common letters at the start and end of the *actual* sequence, as opposed to $E_{\text{fail}}^{\text{com}}$ in the recursion relation in Eq. (3.127), which deals with the common letters at the start of the actual (or failed) sequence and at the end of the failed sequence.

As an example, let's verify that Eq. (3.130) reproduces the $E_{\text{HTHHH}} = 34$ result we obtained in the preceding remark. Since HTHHH has only an H in common at the start and end, Eq. (3.130) gives $E_{\text{HTHHH}} = 2^5 + E_{\text{H}} = 32 + 2 = 34$, as desired. As another example, the 7-letter sequence HTHHHTH has an HTH in common at the start and end, so we obtain $E_{\text{HTHHHTH}} = 2^7 + E_{\text{HTH}} = 128 + 10 = 138$. If you want to verify this with the recursion relation in Eq. (3.127), you can show that the E_{sequence} values for the increasing internal sub-sequences (H, HT, HTH, HTHH, etc.) are, respectively, 2, 4, 10, 18, 34, 68, and finally 138, as desired. (Much of this work was already done in the preceding remark.)

As an exercise, you can show that for a given length n of a sequence, the largest possible E_{sequence} value is $2^{n+1} - 2$, obtained when all n letters are the same.

There must be a clean way to directly prove that Eq. (3.127) implies Eq. (3.130), but it eludes me. ♣

30. **Staying ahead**

Consider a two-dimensional lattice in which a vote for A is signified by a unit step in the positive x-direction, and a vote for B is signified by a unit step in the positive y-direction. The counting of the votes until the final tally (where A has a votes and B has b votes) corresponds to a path from the origin to the point (a, b), with $a \geq b$. There are $\binom{a+b}{a}$ (or equivalently $\binom{a+b}{b}$) such paths, because any a steps of the total $a + b$ steps can be chosen to be the ones in the x-direction. All of these paths from the origin to (a, b) are equally likely, as you can show.[12] The probability that a particular path corresponds to the way the votes are counted is therefore $1/\binom{a+b}{a}$.

---

[11]For an n-letter sequence, the first n letters are of course the same as the last n letters, since both of these sets are the entire sequence. But it is understood that E_{com} refers to at most $n - 1$ letters, because we are concerned with the possible head start in obtaining the next sequence, which means that we can use at most $n - 1$ letters from the original sequence. So, for example, in HHH we have $E_{\text{com}} = E_{\text{HH}}$, which you can show equals 6.

[12]The probabilities are all equal to $a!b!/(a + b)!$, as you can see by picking an arbitrary path and writing down the product of the probabilities of each step (which depend on the numbers of uncounted votes for A and B). This result can be written as $1/\binom{a+b}{a}$, in agreement with the fact that $\binom{a+b}{a}$ is the total number of paths.

If A's sub-total is always greater than or equal to B's sub-total, then the path always remains in the $x \geq y$ (lower-right) part of the plane. The problem can therefore be solved by finding the number N_g of paths that reach the point (a, b) without passing through the $y > x$ (upper-left) region. (We'll call these the "good" paths; hence the subscript "g.") It will actually be easier to find the number N_b of paths that reach the point (a, b) and that do pass through the $y > x$ region. (We'll call these the "bad" paths.) The desired probability that A's sub-total is always greater than or equal to B's sub-total is then equal to $1 - N_b / \binom{a+b}{a}$.

Claim: *The number of "bad" paths from the origin to (a, b) (that is, the number of paths that pass through the $y > x$ region) equals $N_b = \binom{a+b}{b-1}$.*

Proof: The first unit step is to either the point $(0, 1)$ or the point $(1, 0)$. So the number of bad paths from $(0, 0)$ to (a, b) equals the number of bad paths from $(0, 1)$ to (a, b) plus the number of bad paths from $(1, 0)$ to (a, b). Let's look at these two classes of bad paths.

- Since $(0, 1)$ is in the $y > x$ region, the number of bad paths from $(0, 1)$ to (a, b) is simply all of the paths from $(0, 1)$ to (a, b), which equals $\binom{a+(b-1)}{b-1}$.
- We claim that the number of bad paths from $(1, 0)$ to (a, b) equals the total number of paths from $(-1, 2)$ to (a, b). This follows from the fact that any bad path from $(1, 0)$ to (a, b) must proceed via a point on the line $y = x + 1$ (which is the start of the $y > x$ region), as shown in Fig. 3.44. This implies that for every bad path from $(1, 0)$ to (a, b), there is a corresponding path from $(-1, 2)$ to (a, b). This correspondence is obtained by reflecting, in the line $y = x + 1$, the portion of the bad path between $(1, 0)$ and the first point of contact with the line $y = x + 1$. An example of such a reflection is shown in Fig. 3.44. Conversely, for every path from $(-1, 2)$ to (a, b), there is a corresponding bad path from $(1, 0)$ to (a, b), obtained by the reflecting the portion of the path between $(-1, 2)$ and the first point of contact with the line $y = x + 1$.[13] There is therefore a one-to-one correspondence between the *bad* paths starting at $(1, 0)$ and *all* of the paths starting at $(-1, 2)$. The total number of paths from $(-1, 2)$ to (a, b) is $\binom{(a+1)+(b-2)}{b-2} = \binom{a+b-1}{b-2}$, so this is the number of bad paths from $(1, 0)$ to (a, b).[14]

Combining the above two classes of bad paths, we see that the total number of bad paths from $(0, 0)$ to (a, b) is

$$N_b = \binom{a+b-1}{b-1} + \binom{a+b-1}{b-2} = \binom{a+b}{b-1}, \tag{3.131}$$

where we have used the fact that a given entry in Pascal's triangle (which is a particular binomial coefficient) equals the sum of the two entries above it.

---

[13]This reasoning holds only if every path from $(-1, 2)$ to (a, b) actually does touch the line $y = x + 1$, so that there is indeed a first point of contact. This requires that $a \geq b - 1$, which is satisfied here because we are assuming $a \geq b$.

[14]This result holds only if $b \geq 2$, because otherwise the lower entry in the binomial coefficient is negative. But if $b = 0$ or 1, then every path from $(1, 0)$ to (a, b) is good, consistent with the fact that there are no paths from $(-1, 2)$ to $(a, 0)$ or $(a, 1)$; steps in the negative y-direction aren't allowed.

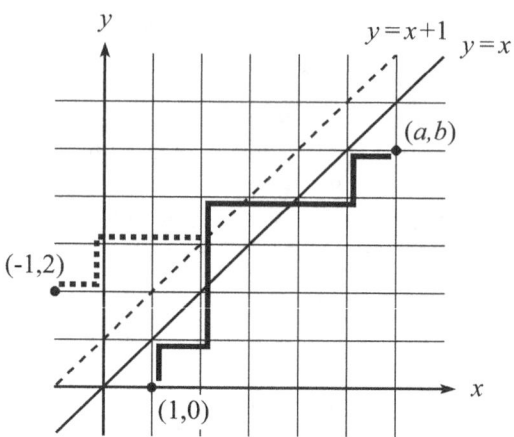

Figure 3.44

Alternatively, you can just write out the binomial coefficients on both sides of Eq. (3.131) in terms of factorials, and then demonstrate the equality. ∎

The probability that A's sub-total is always greater than or equal to B's sub-total is therefore

$$P_{A \geq B} = 1 - N_b \bigg/ \binom{a+b}{a} = 1 - \binom{a+b}{b-1} \bigg/ \binom{a+b}{a}$$
$$= 1 - \frac{(a+b)!}{(b-1)!(a+1)!} \bigg/ \frac{(a+b)!}{a!b!} = 1 - \frac{b}{a+1}. \quad (3.132)$$

If $b = a + 1$ then $P_{A \geq B} = 0$. This makes sense, because if b is larger than a, there is no way for A to always be ahead of B in the counting, since B is ahead at least at the very end. A mentioned in Footnote 13, the derivation of Eq. (3.132) is valid only if $a \geq b - 1 \iff b \leq a + 1$. But for larger values of b (and even for $b = a + 1$), $P_{A \geq B} = 0$ anyway, because there is no way for A to always be ahead of B.

If $b = a$ (the election is a tie), then Eq. (3.132) gives $P_{A \geq B} = 1/(a+1)$. For large a, this goes to zero, which makes sense; a long 50-50 random walk is very unlikely to always remain on one side of (including at) the origin. (Problem 31 discusses various aspects of random walks.) Technically, the ballot counting in the $a = b$ case isn't a true 50-50 random walk, because it is constrained to end up at a given point, namely (a, a). But for large a, the counting is essentially a 50-50 random walk, for the present purposes.

For large but general values of a and b, we can ignore the 1 in the denominator in Eq. (3.132), so we obtain $P_{A \geq B} = 1 - b/a$. Another way to state this result is that b/a is the probability that B is ahead of A at some point in the counting process. So if, for example, B receives one million votes and A receives two million, then there is (essentially) a 1/2 chance that B is ahead at some point in the counting. As an exercise, you can also derive this (very clean) probability of b/a by using the method in the first solution of Problem 28. Hint: The differences between neighboring $P(k)$'s (with $P(k)$ appropriately defined) now form a geometric progression instead of being equal, as they were in the third line following Eq. (3.109).

VARIATION: What is the probability that A's sub-total is always strictly greater than B's? (Ignore the initial zero-zero tie.) Try to solve this before reading further.

The same "reflection" reasoning holds, except that now we must reflect across the line $y = x$, because any path that touches the line $y = x$ is now "bad." The numbers of paths in the two classes of bad paths in the above proof are now the same; both classes involve paths that go from $(0, 1)$ to (a, b), because the reflection of the point $(1, 0)$ in the line $y = x$ is the point $(0, 1)$. Therefore, the number N_b of bad paths is twice the number of paths from $(0, 1)$ to (a, b), which gives $N_b = 2\binom{a+b-1}{b-1}$. The probability that A's sub-total is always strictly greater than B's sub-total is therefore

$$P_{A>B} = 1 - N_b \bigg/ \binom{a+b}{a} = 1 - 2\binom{a+b-1}{b-1} \bigg/ \binom{a+b}{b}$$

$$= 1 - 2\frac{(a+b-1)!}{(b-1)!a!} \bigg/ \frac{(a+b)!}{a!b!} = 1 - \frac{2b}{a+b} = \frac{a-b}{a+b}. \qquad (3.133)$$

If $a = b$ this equals zero as it should, because at least at the very end, A isn't strictly ahead of B.

31. **Random walk**

    (a) We can list out a given sequence of $2n$ steps by using the labels R or L for a right or left step, for example, RLLRLRRR.... There are two choices for each of the $2n$ steps, so there are 2^{2n} possible sequences with length $2n$. Every sequence occurs with the same probability of $(1/2)^{2n}$, because each letter (R or L) for each step occurs with probability $1/2$. The desired probability p_{2n} therefore equals $1/2^{2n}$ times the number of sequences that have equal numbers of R's and L's (the condition to get back to the origin). This number equals $\binom{2n}{n}$ because there are $\binom{2n}{n}$ ways to choose which n of the $2n$ steps in the sequence we label with an R (or an L). Therefore,

$$p_{2n} = \frac{1}{2^{2n}} \binom{2n}{n}. \qquad (3.134)$$

This is also the answer to the question: What is the probability of getting n Heads in $2n$ coin flips? This coin-flip setup is equivalent to the 1-D random-walk setup, because the direction of each step can be determined by a coin flip.

REMARK: When n is large, Stirling's formula from Problem 52 leads to a nice approximation to the result in Eq. (3.134). Using $n! \approx n^n e^{-n} \sqrt{2\pi n}$, we have

$$p_{2n} = \frac{1}{2^{2n}}\binom{2n}{n} = \frac{1}{2^{2n}}\frac{(2n)!}{n!n!} \approx \frac{1}{2^{2n}}\frac{(2n)^{2n}e^{-2n}\sqrt{2\pi \cdot 2n}}{\left(n^n e^{-n}\sqrt{2\pi n}\right)^2}. \quad (3.135)$$

All of the factors except the square roots conveniently cancel, so we're left with

$$p_{2n} \approx \frac{\sqrt{2\pi \cdot 2n}}{2\pi n} = \frac{1}{\sqrt{\pi n}}, \quad (3.136)$$

which is about as simple a result as we could hope for. As a few examples, we find that the probability of getting 50 Heads in 100 coin flips is approximately $1/\sqrt{\pi \cdot 50} \approx 8\%$, and the probability of getting 500 Heads in 1000 coin flips is approximately $1/\sqrt{\pi \cdot 500} \approx 2.5\%$. These approximations are very good; for the $n = 50$ case, the exact result in Eq. (3.134) gives 0.07959, while the $1/\sqrt{\pi \cdot 50}$ approximation gives 0.07979, which is good to 0.25%. The accuracy gets even better as n increases. ♣

(b) We can associate our 1-D random walk with a walk in the 2-D x-y plane, by identifying a rightward step with a unit step in the positive x-direction, and a leftward step with a unit step in the positive y-direction. A sequence of R and L letters with length $2n$ describing a 1-D walk can now be interpreted as a path with length $2n$ in the x-y plane. All points on the $y = x$ line correspond to the origin of the 1-D walk.

Let G_{2n} (with "G" for "good") be the number of paths that return to the origin for the first time after $2n$ steps. (Whenever we use the word "after" in this problem, we mean *right* after.) Since the total number of paths with length $2n$ is 2^{2n}, the desired probability is $f_{2n} = G_{2n}/2^{2n}$. G_{2n} equals the total number of paths to the point (n, n) (which is $\binom{2n}{n}$) minus the number, B_{2n} (with "B" for "bad"), of paths to (n, n) that return to the origin at some time before $2n$ steps. That is, $G_{2n} = \binom{2n}{n} - B_{2n}$. Our goal is therefore to determine B_{2n}. We can do this as follows.

In Fig. 3.45, a path from the origin to (n, n) (we've chosen $n = 6$ for concreteness) must involve going from either A to C, or A to D, or B to C, or B to D. The A-to-D and B-to-C types of paths necessarily cross the $y = x$ line, so they necessarily return to the origin at some time before $2n$ steps. They are therefore all bad paths. There are $\binom{2n-2}{n}$ paths of each of these types, because they involve n steps in one direction and $n - 2$ in the other. So we have $2 \cdot \binom{2n-2}{n}$ bad paths (so far).

For the A-to-C and B-to-D types of paths, some of these touch the $y = x$ line, and some don't. How many do? That is, how many are bad paths? We can answer this using the reflection technique from Problem 30. Consider a bad path from A to C (one that touches the $y = x$ line). There is a one-to-one correspondence between the bad paths going from A to C and *all* of the paths going from B to C. This correspondence is obtained by reflecting, in the $y = x$ line, the portion of the bad path between A and the first point of contact with the $y = x$ line. This reflection turns the starting point at $A = (1, 0)$ into the starting point at $B = (0, 1)$. The number of A-to-C paths

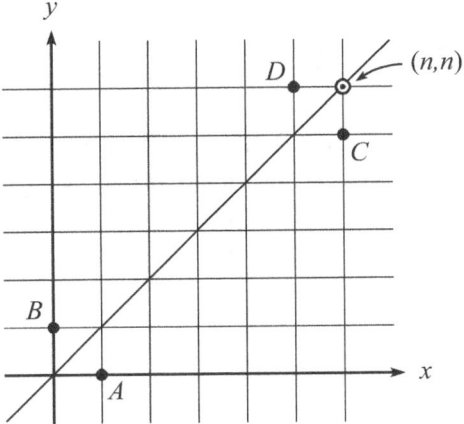

Figure 3.45

that are bad is therefore the same as the total number of *B*-to-*C* paths, which is just the $\binom{2n-2}{n}$ result in the preceding paragraph. Likewise for the subset of the *B*-to-*D* paths that are bad. So we have another $2 \cdot \binom{2n-2}{n}$ bad paths. The total number of bad paths from the origin to (n, n) (paths that return to the origin at some time before $2n$ steps) is therefore $B_{2n} = 4 \cdot \binom{2n-2}{n}$. The number of good paths (ones that return to the origin for the first time after $2n$ steps) is then

$$G_{2n} = \binom{2n}{n} - B_{2n} = \binom{2n}{n} - 4\binom{2n-2}{n}. \quad (3.137)$$

The desired probability of returning to the origin for the first time after $2n$ steps is therefore

$$f_{2n} = \frac{G_{2n}}{2^{2n}} = \frac{1}{2^{2n}}\left(\binom{2n}{n} - 4\binom{2n-2}{n}\right). \quad (3.138)$$

Simplifying this gives

$$\begin{aligned}f_{2n} &= \frac{1}{2^{2n}}\left(\frac{(2n)!}{n!n!} - 4\frac{(2n-2)!}{n!(n-2)!}\right) \\ &= \frac{1}{2^{2n}}\frac{(2n-2)!}{n!n!}\Big((2n)(2n-1) - 4n(n-1)\Big) \\ &= \frac{1}{2^{2n-1}n}\binom{2n-2}{n-1}. \quad (3.139)\end{aligned}$$

As an exercise, you can combine this result with the one for p_{2n} in Eq. (3.134) to quickly show that $f_{2n}/p_{2n} = 1/(2n-1)$. In other words, given that you have returned to the origin after $2n$ steps, there is a $1/(2n-1)$ chance that this is your first return to the origin. As a double check, this ratio equals 1 when $n = 1$. And it equals $1/3$ when $n = 2$, which you can verify is correct.

For future reference, note that Eq. (3.138) can be written as

$$f_{2n} = \frac{1}{2^{2n}}\binom{2n}{n} - \frac{1}{2^{2n-2}}\binom{2n-2}{n} = P_{(n,n)} - P_{(n,n-2)}, \quad (3.140)$$

where $p_{(a,b)}$ is the probability of ending up (not necessarily for the first time) at the point (a, b) (after $a + b$ steps, of course). $p_{(n,n)}$ is simply what we've been calling p_{2n}.

(c) Let's calculate $p_{2n-2} - p_{2n}$ and show that it equals the result for f_{2n} in Eq. (3.139). We have

$$\begin{aligned}p_{2n-2} - p_{2n} &= \frac{1}{2^{2n-2}}\binom{2n-2}{n-1} - \frac{1}{2^{2n}}\binom{2n}{n}\\ &= \frac{1}{2^{2n-2}}\frac{(2n-2)!}{(n-1)!(n-1)!} - \frac{1}{2^{2n}}\frac{(2n)!}{n!n!}\\ &= \frac{1}{2^{2n}}\frac{(2n-2)!}{n!n!}\left(4n^2 - (2n)(2n-1)\right)\\ &= \frac{1}{2^{2n-1}n}\binom{2n-2}{n-1},\end{aligned} \quad (3.141)$$

in agreement with the f_{2n} in Eq. (3.139). A quick corollary of this $f_{2n} = p_{2n-2} - p_{2n}$ result is that

$$f_2 + f_4 + f_6 \cdots = (p_0 - p_2) + (p_2 - p_4) + (p_4 - p_6) + \cdots . \quad (3.142)$$

This sum telescopes to p_0, which is simply 1. (The probability of being at the origin after zero steps is 1.) So the probability is 1 that you eventually return to the origin for the first time (since that is what the f_{2n}'s represent). In other words, you are guaranteed to eventually return to the origin in a 1-D random walk. More precisely, for any ϵ, there is an n such that the probability of returning to the origin before or on the $(2n)$th step exceeds $1 - \epsilon$.

However, it turns out that the expected number of steps it takes to get back to the origin for the first time (which is $S = \sum_1^\infty 2n \cdot f_{2n}$) is infinite. This is true because this sum equals

$$\begin{aligned}S &= 2f_2 + 4f_4 + 6f_6 + \cdots\\ &= 2(p_0 - p_2) + 4(p_2 - p_4) + 6(p_4 - p_6) + \cdots\\ &= 2(p_0 + p_2 + p_4 + p_6 + \cdots).\end{aligned} \quad (3.143)$$

Since we're just trying to show that this sum diverges, it suffices to use the approximate form of p_{2n} in Eq. (3.136), which tells us that $p_{2n} \propto 1/\sqrt{n}$ for large n. And since the sum of $1/\sqrt{n}$ diverges (because the integral of $1/\sqrt{n}$ diverges), we see that the expected number of steps S diverges.

REMARK: In the $p_{(a,b)}$ notation of Eq. (3.140), the $f_{2n} = p_{2n-2} - p_{2n}$ result can be written as

$$f_{2n} = P_{(n-1,n-1)} - P_{(n,n)}. \quad (3.144)$$

If we equate this expression for f_{2n} with the one in Eq. (3.140), we obtain

$$p_{(n,n)} - p_{(n,n-2)} = p_{(n-1,n-1)} - p_{(n,n)}$$
$$\implies p_{(n,n)} = \frac{1}{2}\left(p_{(n,n-2)} + p_{(n-1,n-1)}\right). \tag{3.145}$$

The intuitive interpretation of this equation is that in order to get to the point (n, n), you must pass through either $(n - 1, n - 1)$ or $(n, n - 2)$ or $(n - 2, n)$ (with the probabilities of passing through the latter two points being equal). From $(n - 1, n - 1)$, there is a $1/2$ chance that you end up at (n, n) two steps later, as you can verify. And from each of $(n, n - 2)$ and $(n - 2, n)$, there is a $1/4$ chance that you end up at (n, n) two steps later. We therefore arrive at Eq. (3.145). ♣

(d) The probability that you *do* return to the origin at some point in a walk with length $2n$ is $f_2 + f_4 + \cdots + f_{2n}$. (We can indeed simply add these probabilities without worrying about double counting any paths, because the f's are associated with outcomes that are mutually exclusive.) The desired probability that you do *not* return to the origin at some point in a walk with length $2n$ is therefore (using $f_{2n} = p_{2n-2} - p_{2n}$, along with $p_0 = 1$)

$$\begin{aligned} a_{2n} &= 1 - (f_2 + f_4 + \cdots + f_{2n}) \\ &= 1 - \left((p_0 - p_2) + (p_2 - p_4) + \cdots + (p_{2n-2} - p_{2n})\right) \\ &= 1 - p_0 + p_{2n} \\ &= p_{2n}, \end{aligned} \tag{3.146}$$

as desired. As a consistency check, note that this $a_{2n} = p_{2n}$ result turns the $f_{2n} = p_{2n-2} - p_{2n}$ result in part (c) into $f_{2n} = a_{2n-2} - a_{2n}$. This is indeed correct, because if you return to the origin for the first time on the $(2n)$th step, then two things must be true: (1) it must be the case that you never returned by the $(2n - 2)$th step (which happens with probability a_{2n-2}), and (2) it must *not* be the case that you never returned by the $(2n)$th step; so we must subtract off the probability a_{2n}.[15] Hence $f_{2n} = a_{2n-2} - a_{2n}$.

REMARK: Here is another (longer) way to show that $a_{2n} = p_{2n}$. It uses the reflection technique from Problem 30. After $2n$ steps, your position will be at one of the 11 large dots (we've chosen $n = 5$ for concreteness) shown in Fig. 3.46. Each one of the 2^{2n} possible paths has the same probability $1/2^{2n}$, so the $a_{2n} = p_{2n}$ statement about probabilities can be recast as a statement about numbers of paths: The total number of paths to the dot at (n, n) (which is $\binom{2n}{n}$) equals the total number of "good" paths (ones that don't touch the $y = x$ line) to the 10 other dots. We can demonstrate this as follows.

---

[15]It is legal to simply subtract a_{2n} from a_{2n-2}, because the events associated with a_{2n} are a subset of the events associated with a_{2n-2}. If you've never returned to the origin by the $(2n)$th step, then you've certainly also never returned by the $(2n - 2)$th step.

116   Chapter 3. Solutions

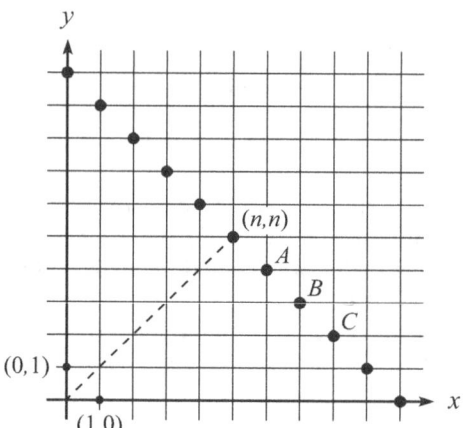

Figure 3.46

Let's first find the number of good paths from the origin to point A. The first unit step must be to the point $(1, 0)$ in order for the path to A to have any chance of being good. So the number of good paths from the origin to A equals the total number of paths from $(1, 0)$ to A (which is $\binom{2n-1}{n-1}$) minus the number of bad paths from $(1, 0)$ to A. (Of the two choices for the lower entry in the binomial coefficient here, we'll use the number of steps in the y-direction.) From the reflection technique in Problem 30 (with the reflection across the $y = x$ line), the number of bad paths from $(1, 0)$ to A equals the total number of paths from $(0, 1)$ to A (which is $\binom{2n-1}{n-2}$). So the number of good paths from the origin to point A is $\binom{2n-1}{n-1} - \binom{2n-1}{n-2}$.

Now consider point B. The same strategy gives the number of good paths from the origin to B as $\binom{2n-1}{n-2} - \binom{2n-1}{n-3}$, as you can verify. Likewise, the number of good paths from the origin to C is $\binom{2n-1}{n-3} - \binom{2n-1}{n-4}$. The pattern continues until the number of good paths to the 2nd-to-last point in the lower-right part of the grid is $\binom{2n-1}{1} - \binom{2n-1}{0}$, and then finally the number of good paths to the last point is just $\binom{2n-1}{0}$. The $\binom{2n-1}{0}$ terms in these last two results correspond to a single horizontal line.

Summing the above results, we see that the total number of good paths to the dots in the lower-right part of the grid is

$$\left[\binom{2n-1}{n-1} - \binom{2n-1}{n-2}\right]$$
$$+ \left[\binom{2n-1}{n-2} - \binom{2n-1}{n-3}\right]$$
$$+ \left[\binom{2n-1}{n-3} - \binom{2n-1}{n-4}\right] + \cdots . \quad (3.147)$$

This sum telescopes to the simple result of $\binom{2n-1}{n-1}$. The same result holds for the dots in the upper-left part of the grid, so the total number of good paths to all the dots (except the one at (n, n)) equals

$$2 \cdot \binom{2n-1}{n-1} = \frac{2(2n-1)!}{(n-1)!n!} = \frac{(2n)(2n-1)!}{n!n!} = \binom{2n}{n}, \qquad (3.148)$$

which agrees with the total number of paths to the dot at (n, n), as desired. As an exercise, you can think about how to generate a 1-to-1 correspondence between the paths to (n, n) and the good paths to the other dots. ♣

32. **Standing in a line**

    FIRST SOLUTION: Let T_N be the expected number of people who are able to make the given statement that they are taller than everyone in front of them. If we consider everyone except the last person in the line (the person who can see everyone else), then this group of $N-1$ people has, by definition, T_{N-1} people (on average) who are able to make the statement. Let us now add on the last person. There is a $1/N$ chance that she is the tallest, in which case she is able to make the statement. We therefore have

    $$T_N = T_{N-1} + \frac{1}{N}. \qquad (3.149)$$

Starting with $T_1 = 1$, we inductively find

$$T_N = 1 + \frac{1}{2} + \frac{1}{3} + \cdots + \frac{1}{N}. \qquad (3.150)$$

For large N, this goes like $\ln N$, which grows very slowly with N.

SECOND SOLUTION: Let T_N be the desired average. Consider the location of the tallest person. If he is the last person in the line (which occurs with probability $1/N$), then the problem reduces to that for the $N-1$ people in front of him. So in this case, we can expect $1 + T_{N-1}$ people who are able to make the given statement.

If the tallest person is the second-to-last person in the line (which occurs with probability $1/N$), then the problem reduces to that for the $N-2$ people in front of him (because the person behind him is not able to make the statement). So in this case, we can expect $1 + T_{N-2}$ people who are able to make the given statement.

Continuing along these lines, and adding up all N possibilities for the location of the tallest person, we find

$$T_N = \frac{1}{N}\Big((1 + T_{N-1}) + (1 + T_{N-2}) + \cdots + (1 + T_1) + (1 + T_0)\Big)$$
$$\implies NT_N = N + T_{N-1} + T_{N-2} + \cdots + T_1. \qquad (3.151)$$

(We have used the fact that $T_0 = 0$, since that scenario involves zero people. But T_0 would cancel out in the following reasoning in any case.) Writing down the analogous equation for $N-1$,

$$(N-1)T_{N-1} = (N-1) + T_{N-2} + T_{N-2} + \cdots + T_1, \qquad (3.152)$$

and then subtracting this from Eq. (3.151), yields

$$NT_N - (N-1)T_{N-1} = 1 + T_{N-1} \implies T_N = T_{N-1} + \frac{1}{N}, \qquad (3.153)$$

which agrees with the recursion relation in the first solution.

33. **Rolling the die**

    To get a feel for the problem, let's work things out for a few small values of N. For $N = 1$, the probability that the first player wins is 1. There is only one possible roll, so it is impossible to beat. For $N = 2$, the probability is 3/4. The first player definitely wins if she rolls a 2, and she has a 1/2 chance of winning if she rolls a 1. Averaging these two cases yields 3/4. And for $N = 3$, the probability is 19/27. The first player definitely wins if she rolls a 3, she has a 2/3 chance of winning if she rolls a 2, and she has a 4/9 chance of winning if she rolls a 1, as you can check. Averaging these three cases yields 19/27. The pattern in these numbers is more evident if we instead look at the probabilities that the first player *loses*. These are 0, 1/4, and 8/27. And if you work things out for $N = 4$, you will obtain 81/256. These probabilities can be written as 0, $(1/2)^2$, $(2/3)^3$, and $(3/4)^4$. We therefore conjecture that the probability, P_L, that the first player *loses* is

    $$P_L = \left(1 - \frac{1}{N}\right)^N. \qquad (3.154)$$

    We'll prove this by proving the following more general claim. Eq. (3.154) is the special case of the claim when $r = 0$.

    Claim: *Let L_r be the probability that a player loses, given that a roll of r has just occurred. Then*

    $$L_r = \left(1 - \frac{1}{N}\right)^{N-r}. \qquad (3.155)$$

    Proof: Assume that a roll of r has just occurred. To determine the probability L_r that the player who goes next loses, let's consider the probability $1 - L_r$ that she wins. In order to win, she must roll a number a greater than r (each of which occurs with probability $1/N$); and her opponent must then lose, given that he needs to beat a roll of a (which occurs with probability L_a). So the probability of winning, given that a roll of r has just occurred, is

    $$1 - L_r = \frac{1}{N}(L_{r+1} + L_{r+2} + \cdots + L_N). \qquad (3.156)$$

    If we write down the analogous equation using $r - 1$ instead of r,

    $$1 - L_{r-1} = \frac{1}{N}(L_r + L_{r+1} + \cdots + L_N), \qquad (3.157)$$

    and then subtract Eq. (3.157) from Eq. (3.156), we obtain

    $$L_{r-1} - L_r = -\frac{1}{N}L_r \implies L_{r-1} = \left(1 - \frac{1}{N}\right)L_r, \qquad (3.158)$$

for all r from 1 to N. Using $L_N = 1$, we find $L_{N-1} = (1 - 1/N)$ and $L_{N-2} = (1 - 1/N)^2$, etc., down to $L_0 = (1 - 1/N)^N$. So in general we have

$$L_r = \left(1 - \frac{1}{N}\right)^{N-r} \qquad (0 \leq r \leq N). \blacksquare \qquad (3.159)$$

Returning to the original problem, we may consider the first player to start out with a roll of $r = 0$ having just occurred. (Having a roll of zero to beat is the same as having no roll to beat.) So the probability P_L that the first player loses is given by $P_L = L_0$. Therefore, the desired probability that the first player wins is

$$P_W = 1 - P_L = 1 - L_0 = 1 - \left(1 - \frac{1}{N}\right)^N. \qquad (3.160)$$

For large N, this probability approaches $1 - 1/e \approx 63.2\%$ (see the first remark below). For a standard die with $N = 6$, P_W equals $1 - (5/6)^6 \approx 66.5\%$.

REMARKS:

1. The fact that $(1 - 1/n)^n$ approaches $1/e$ in the $n \to \infty$ limit is a special case of Eq. (1.5) in Problem 53, with $a = -1/n$. Alternatively, as an exercise you can derive the general relation,

$$\lim_{n \to \infty} \left(1 + \frac{x}{n}\right)^n = e^x, \qquad (3.161)$$

by using the binomial expansion. In the $n \to \infty$ limit, the binomial coefficients simplify to numbers of the form $n^k/k!$. You will end up with the sum $1 + x + x^2/2! + x^3/3! + \cdots$, which is the Taylor series for e^x. (See the appendix for a review of Taylor series.)

2. If we use the identity

$$1 - x^N = (1 - x)(x^{N-1} + x^{N-2} + \cdots + x + 1), \qquad (3.162)$$

then the probability that the first player wins, given in Eq. (3.160), can be written as (letting $x \equiv 1 - 1/N$)

$$1 - \left(1 - \frac{1}{N}\right)^N = \frac{1}{N}\left(\left(1 - \frac{1}{N}\right)^{N-1} + \left(1 - \frac{1}{N}\right)^{N-2} + \cdots + \left(1 - \frac{1}{N}\right)^1 + 1\right). \qquad (3.163)$$

The righthand side shows (using Eq. (3.159)) explicitly the probabilities that the first player wins, depending on what her first roll is. For example, the first term on the righthand side is the probability $1/N$ that the first player rolls a 1, times the probability $(1 - 1/N)^{N-1}$ that the second player loses given that he must beat a 1. ♣

34. **Strands of spaghetti**

Imagine picking the first pair of ends in succession instead of grabbing them simultaneously; this doesn't affect the process. After you have reached into the bowl and pulled out one end, there are $2N - 1$ free ends left in the bowl. When you pick one of these ends, there is a $1/(2N - 1)$ chance of choosing the other end of the strand that you are holding, in which case a loop is formed. There is a $(2N - 2)/(2N - 1)$ chance of choosing one of the $2N - 2$ ends belonging to the other $N - 1$ strands, in which case a loop is not formed. In the former case, you end up with one loop and $N - 1$ remaining strands. In the latter case, you just end up with $N - 1$ strands, because you have simply created a strand with twice the original length, and the length of a strand is irrelevant in this problem.

Therefore, after the first step, we see that no matter what happens, you end up with $N - 1$ strands along with, on average, $1/(2N - 1)$ loops. We can now repeat this reasoning with $N - 1$ strands. After the second step, we are guaranteed to be left with $N - 2$ strands along with, on average, another $1/(2(N-1)-1) = 1/(2N-3)$ loops. This process continues until we are left with one strand, whereupon the final Nth step leaves us with zero strands, and we gain one more loop.

Adding up the average number of loops gained at each stage, we obtain an average total number of loops equal to

$$n = \frac{1}{2N - 1} + \frac{1}{2N - 3} + \cdots + \frac{1}{3} + 1. \tag{3.164}$$

This grows very slowly with N. It turns out that we need $N = 8$ strands in order to expect at least two loops. If we use the ordered pair (n, N) to signify that N strands are needed in order to expect n loops, you can shown numerically that the first few ordered pairs are $(1, 1)$, $(2, 8)$, $(3, 57)$, $(4, 419)$, and $(5, 3092)$. The largeness of these N values is quite surprising. Most people would probably expect far more than five loops to be formed, given 3000 strands of spaghetti.

REMARK: For large N, we can say that the average number n of loops given in Eq. (3.164) is roughly equal to $1/2$ times the sum of the integer reciprocals up to $1/N$. So it approximately equals $(\ln N)/2$. To get a better approximation, let S_N denote the sum of the integer reciprocals up to $1/N$. Then we have (using $S_N \approx \ln N + \gamma$, where $\gamma \approx 0.5772$ is Euler's constant)

$$n + \left(\frac{1}{2} + \frac{1}{4} + \cdots + \frac{1}{2N - 2} + \frac{1}{2N}\right) = S_{2N}$$

$$\implies n + \frac{1}{2}S_N = S_{2N}$$

$$\implies n + \frac{1}{2}\left(\ln N + \gamma\right) \approx \ln(2N) + \gamma$$

$$\implies n \approx \frac{1}{2}\left(\ln N + \gamma + 2\ln 2\right)$$

$$\implies N \approx \frac{e^{2n-\gamma}}{4}. \tag{3.165}$$

You can show that this relation between n and N agrees well with the above numerical results (even though we have no right to expect it to work for these small-N cases). ♣

35. **How much change?**

If the item costs between $N/2$ and N dollars, then you can buy only one item. These two bounds produce remainders of $N/2$ (or technically an infinitesimal amount less than $N/2$) and 0, respectively. The average amount of money left over in this interval of prices is therefore $N/4$. The length of this interval is $N(1 - 1/2) = N/2$, so the probability of the price lying in this interval is $1/2$.

Similarly, if the item costs between $N/3$ and $N/2$ dollars, then you can buy only two items. These two bounds produce remainders of $N/3$ and 0, respectively. The average amount of money left over in this interval is therefore $N/6$. And the probability of the price lying in this interval is $1/2 - 1/3 = 1/6$.

Continuing in this manner, we see that if the item costs between $N/(n+1)$ and N/n, then you can buy only n items. These two bounds produce remainders of $N/(n+1)$ and 0, respectively. The average amount of money left over in this interval is therefore $N/(2(n+1))$. And the probability of the price lying in this interval is $1/n - 1/(n+1) = 1/n(n+1)$. (The expression on the lefthand side of this equation will be the more useful one in the sum below.)

If we add up the average amount of money left over in the various intervals, weighted by the probability of being in each interval, we find that the expected amount of money M left over is

$$M = \sum_{n=1}^{\infty} \left(\frac{1}{n} - \frac{1}{n+1}\right) \frac{N}{2(n+1)}$$
$$= \frac{N}{2} \sum_{n=1}^{\infty} \left(\frac{1}{n(n+1)} - \frac{1}{(n+1)^2}\right)$$
$$= \frac{N}{2} \sum_{n=1}^{\infty} \left(\left[\frac{1}{n} - \frac{1}{n+1}\right] - \frac{1}{(n+1)^2}\right)$$
$$= \frac{N}{2}\left(1 - \left(\frac{\pi^2}{6} - 1\right)\right)$$
$$= N\left(1 - \frac{\pi^2}{12}\right). \tag{3.166}$$

In the third line, we used the fact that the sum in brackets telescopes to 1, and also that $\sum_{n=1}^{\infty} 1/k^2 = \pi^2/6$. (Our sum starts at $k = 2$.) Since $\pi^2/12 \approx 0.82$, the average amount of money left over is roughly $(0.18)N$ dollars. Note that what we have essentially done in this problem is find the area under the sawtooth "curve" in Fig. 3.47.

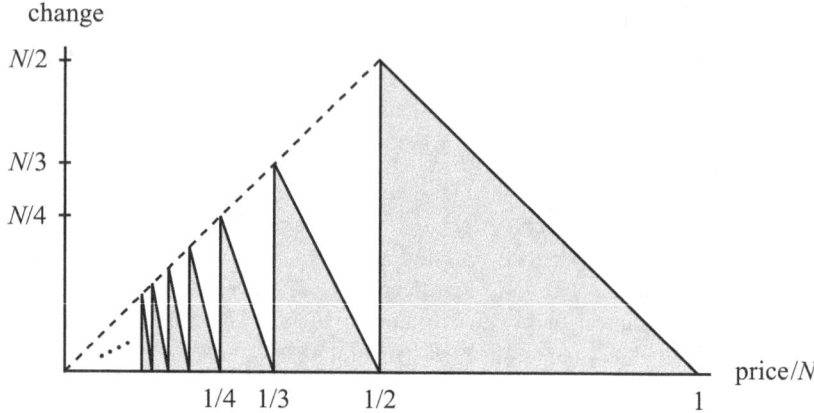

Figure 3.47

36. Relatively prime numbers

Two numbers are relatively prime if they have no common factor, which is the case if and only if they have no common prime factor. Now, the probability that two random numbers both have a given prime p as a factor is $1/p^2$. So the probability that they do *not* have p as a common factor is $1 - 1/p^2$. The probability that two numbers have *no* common prime factors (that is, the probability that the two numbers are relatively prime) is therefore

$$P = (1 - 1/2^2)(1 - 1/3^2)(1 - 1/5^2)(1 - 1/7^2)(1 - 1/11^2) \cdots . \qquad (3.167)$$

Using

$$\frac{1}{1-x} = 1 + x + x^2 + x^3 + \cdots, \qquad (3.168)$$

P can be rewritten as

$$P = \Big(\big(1 + 1/2^2 + 1/2^4 + \cdots\big)\big(1 + 1/3^2 + 1/3^4 + \cdots\big) \cdots \Big)^{-1}. \qquad (3.169)$$

By the Unique Factorization Theorem (every positive integer, except 1, is expressible as the product of primes in exactly one way), we see that the above product will generate every number of the form $1/n^2$, where n is positive integer. For example, $1/360^2$ comes from the product $(1/2^6)(1/3^4)(1/5^2)$. The above expression for P is therefore equivalent to

$$P = \big(1 + 1/2^2 + 1/3^2 + 1/4^2 + 1/5^2 + 1/6^2 + \cdots\big)^{-1}. \qquad (3.170)$$

And since the sum of the reciprocals of the squares of all of the positive integers is known to be $\pi^2/6$,[16] the desired probability is $P = 6/\pi^2 \approx 61\%$.

---

[16] The task of calculating the sum of the reciprocals of the squares is known as the Basel problem, and it has many solutions you can look up.

REMARKS:

1. The probability that n random numbers all have a given prime p as a factor is $1/p^n$. So the probability that they do *not* all have p as a common factor is $1 - 1/p^n$. In exactly the same manner as above, we find that the probability P_n that n numbers have no common factor among all of them is

$$P_n = (1 + 1/2^n + 1/3^n + 1/4^n + 1/5^n + 1/6^n + \cdots)^{-1}. \qquad (3.171)$$

The sum inside the parentheses is, by definition, the Riemann zeta function, $\zeta(n)$. It can be calculated exactly for even values of n, but only numerically for odd values. (Well, except for $n = 1$, where the sum is infinite.) For $n = 2$, we have our original $P \equiv P_2 = 6/\pi^2$. For $n = 4$, the known value $\zeta(4) = \pi^4/90$ tells us that the probability that four random numbers do not all have a common factor is $P_4 = 90/\pi^4 \approx 92\%$.

2. We can also perform the somewhat silly exercise of applying the above procedure to the case of $n = 1$. For $n = 1$ the question becomes: What is the probability P_1 that a randomly chosen positive integer does not have a factor? Well, 1 is the only positive integer without any factors, so the probability is $1/\infty = 0$. And indeed,

$$\begin{aligned} P_1 &= (1 - 1/2)(1 - 1/3)(1 - 1/5)(1 - 1/7) \cdots \\ &= (1 + 1/2 + 1/3 + 1/4 + 1/5 + 1/6 \cdots)^{-1} \\ &= 1/\infty, \end{aligned} \qquad (3.172)$$

because the sum of the reciprocals of all of the positive integers is infinite.

3. Let $\phi(n)$ equal the number of integers less than n that are relatively prime to n. Then $\phi(n)/n$ equals the probability that a randomly chosen integer is relatively prime to n. (This is true because any integer is relatively prime to n if and only if its remainder, when divided by n, is relatively prime to n.) The result of our original problem therefore tells us that the average value of $\phi(n)/n$ is $6/\pi^2$. In other words, $(1/N) \sum_{n=1}^{N} \phi(n)/n$ approaches $6/\pi^2$ as $N \to \infty$. You can verify this numerically with Mathematica.

4. To be precise about what we mean by probabilities in this problem, we should word the question as: Let N be a very large integer. Pick two random integers less than or equal to N. What is the probability that these integers are relatively prime, in the $N \to \infty$ limit? The solution would then be slightly modified, in that the relevant primes p would be cut off at N, and "edge effects" due to the finite size of N would have to be considered. (If N isn't a multiple of p, then the probability that an integer (less than or equal to N) is divisible by p isn't exactly equal to $1/p$.) But these effects become negligible in the $N \to \infty$ limit. This is true because edge effects are negligible for small primes. And small primes are the only ones that matter, because large primes contribute negligibly to Eq. (3.167); truncating the primes in Eq. (3.167) even at just 17 yields $P \approx 0.616$, which is very close to the actual answer of $P = 6/\pi^2 \approx 0.608$. ♣

37. **The hotel problem**

In figuring out the probability of success (choosing the cheapest hotel) when applying the given strategy, it is helpful to organize the different cases according to what the highest-ranking hotel is (in order of cheapness) in the first fraction x. Let H_1 denote the cheapest hotel, H_2 the second cheapest, etc.

Assume that H_1 is among the first fraction x, which happens with probability x. (Technically this isn't true unless Nx is an integer. But for large N, which we are assuming, we don't need to worry about this distinction.) In this case there is guaranteed failure, because you will pass up this hotel when applying the given strategy of passing on the first fraction x.

Assume that H_2 is the cheapest hotel among the first x, which happens with probability $x(1-x)$; this is the probability that H_2 is in the first x, times the probability that H_1 is not.[17] In this case there is guaranteed success, because you will choose H_1 when you encounter it, according to the given strategy.

Assume that H_3 is the cheapest hotel among the first x, which happens with probability $x(1-x)^2$; this is the probability that H_3 is in the first x, times the probability that H_2 is not, times the probability that H_1 also is not (again, see the remark below). In this case, you have success $1/2$ of the time, because there is a $1/2$ chance that H_1 comes before H_2. (If H_2 comes first then you will choose it, according to the given strategy.)

Continuing in this fashion, we see that the probability of success, P, is

$$P(x) = 0 + x(1-x) + \frac{1}{2}x(1-x)^2 + \frac{1}{3}x(1-x)^3 + \cdots$$

$$= \sum_{k=1}^{\infty} \frac{1}{k} x(1-x)^k. \tag{3.173}$$

The $1/k$ factor comes from the probability that H_1 is first among the top k hotels, all of which are assumed to lie in the final $(1-x)$ fraction. We can write Eq. (3.173) in closed form by using the Taylor series $\ln(1-y) = -(y + y^2/2 + y^3/3 + \cdots)$, with $y = 1-x$. (See the appendix for a review of Taylor series.) This gives

$$P(x) = -x \ln x. \tag{3.174}$$

Setting the derivative of this equal to zero to find the maximum gives $-(1 + \ln x) = 0 \implies x = 1/e$. The associated value of P is $P(1/e) = -(1/e)\ln(1/e) = 1/e$. Therefore, when applying the given strategy, you want to pass up on $1/e \approx 37\%$ of the hotels, and then pick the next one that is cheaper than all the ones you've seen. Your probability of getting the cheapest one is then $1/e \approx 37\%$. (It's always nice when an answer involves e!)

REMARK: Concerning Footnote 17: For sufficiently large N, the actual probabilities are arbitrarily close to the $x(1-x)^k/k$ probabilities we used in Eq. (3.173),

---

[17]The $(1-x)$ factor technically isn't correct, because there are only $N-1$ spots available for H_1, given that H_2 has been placed. So the probability that H_1 is not in the first x is actually $(N - Nx)/(N-1)$. But for large N, the "1" term is negligible, so we obtain $1-x$. See the remark at the end of the solution.

for small values of k. And small values of k are the only ones we are concerned with, because successive terms in Eq. (3.173) are suppressed by at least a factor of $(1 - 1/e)$. The terms therefore become negligibly small at a k value that is independent of N. ♣

38. **Decreasing numbers**

    FIRST SOLUTION: Imagine picking a large set of numbers (randomly distributed between 0 and 1) in succession and listing them out. For the present purposes, pay no attention to the relative sizes; keep picking numbers even if a number is greater than the previous one. Label this long sequence of numbers as $x_1, x_2, x_3, \ldots$ in the order you picked them.

    There is a $p_2 = 1/2$ chance that $x_1 > x_2$ (because each of these two numbers is equally likely to be the larger one). And there is a $p_3 = 1/3!$ chance that $x_1 > x_2 > x_3$ (because the 3! possibilities of the ordered ranking of these three numbers are all equally likely). Likewise, there is a $p_4 = 1/4!$ chance that $x_1 > x_2 > x_3 > x_4$. And so on.

    Now back to the original game, where you stop picking when you obtain a number greater than the previous one. You will necessarily pick at least two numbers. The probability that you pick exactly two is equal to the probability that $x_1 < x_2$, which is $1 - p_2 = 1/2$.

    If the game lasts exactly three picks, then two things must happen: We must have $x_1 > x_2$ (so that the game continues to the third pick), and we must also have $x_2 < x_3$ (so that the game stops after the third pick). The probability of both of these things happening equals the probability that $x_1 > x_2$ minus the probability that $x_1 > x_2 > x_3$. That is, the probability equals $p_2 - p_3$.

    Similarly, the game lasts exactly four picks if $x_1 > x_2 > x_3$ and $x_3 < x_4$. The probability of both of these things happening equals the probability that $x_1 > x_2 > x_3$ minus the probability that $x_1 > x_2 > x_3 > x_4$. That is, the probability equals $p_3 - p_4$.

    Continuing in this manner, we find that the expected total number T of picks is

    $$\begin{aligned} T &= 2(1 - p_2) + 3(p_2 - p_3) + 4(p_3 - p_4) + \cdots \\ &= 2 + p_2 + p_3 + p_4 + \cdots \\ &= 1 + 1 + \frac{1}{2!} + \frac{1}{3!} + \frac{1}{4!} + \cdots \\ &= e \approx 2.718. \end{aligned} \qquad (3.175)$$

    Could the answer really have been anything else?

    SECOND SOLUTION: Let $E(x)$ be the expected number of numbers you have yet to pick, given that you have just picked the number x. Then, for example, $E(0) = 1$, because the next number you pick is guaranteed to be greater than $x = 0$, whereupon the game stops. The desired expected total number T of picks in the game is simply $T = E(1)$, because the first pick is automatically less than 1, so the number of picks *after* starting a game with the number 1 is equal to the total number of picks in a game starting with a random number. Let's calculate $E(x)$.

Imagine picking the next number, having just picked x. There is a $1 - x$ chance that this next number is greater than x, in which case the game stops. So in this case it takes you just one pick after the number x. If, on the other hand, you pick a number y that is less than x, then you can expect to pick $E(y)$ numbers after that. So in this case it takes you an average of $E(y) + 1$ total picks after the number x. The probability of picking a number in a range dy around y is simply dy, so the preceding two scenarios yield the relation,

$$E(x) = 1 \cdot (1 - x) + \int_0^x (E(y) + 1)\, dy$$
$$= 1 + \int_0^x E(y)\, dy. \tag{3.176}$$

Differentiating this with respect to x (and using the fundamental theorem of calculus) gives $E'(x) = E(x)$, which means that E must be an exponential function: $E(x) = Ae^x$, where A is some constant. If you want to be rigorous, you can separate variables and integrate:

$$\frac{dE}{dx} = E \implies \int \frac{dE}{E} = \int dx \implies \ln E = x + C \implies E(x) = Ae^x, \tag{3.177}$$

where $A \equiv e^C$. The condition $E(0) = 1$ yields $A = 1$. Hence

$$E(x) = e^x. \tag{3.178}$$

As mentioned above, the expected total number of picks is $T = E(1)$. Therefore, since $E(1) = e$ we have

$$T = e. \tag{3.179}$$

THIRD SOLUTION: Let $p(x)\, dx$ be the probability that a number between x and $x + dx$ is picked as part of a decreasing sequence. By this probability we mean: Play the game a million times, and count the number of times a number between x and $x + dx$ appears (excluding the last pick, which is an increase), and then divide by a million. We can find $p(x)$ by adding up the probabilities, $p_j(x)\, dx$, that a number between x and $x + dx$ is picked on the jth pick of a decreasing sequence.

To determine the various $p_j(x)$ values, imagine picking a large set of numbers (randomly distributed between 0 and 1) in succession and listing them out, as we did in the first solution above. Pay no attention to the relative sizes; it's fine if a number is larger than the previous one. Let's call these sequences *general* ones, and let's call the monotonically decreasing sequences that we're interested in *decreasing* ones (naturally). Consider the first few values of j:

- The probability that a number between x and $x + dx$ is picked first in a decreasing (or general, too) sequence is simply dx.

- The probability that a number between x and $x + dx$ is picked second in a decreasing sequence is $(1 - x)\,dx$, because dx is the probability that we pick such a number on the second pick in a general sequence, and $1 - x$ is the probability that the first number is greater than x (thereby making the sequence a decreasing one).
- The probability that a number between x and $x + dx$ is picked third in a decreasing sequence is $(1/2)(1-x)^2\,dx$, because dx is the probability that we pick such a number on the third pick in a general sequence, and $(1-x)^2$ is the probability that the first two numbers are both greater than x (necessary for a decreasing sequence), and furthermore $1/2$ is the probability that these numbers are picked in decreasing order (also necessary for a decreasing sequence).
- The probability that a number between x and $x + dx$ is picked fourth in a decreasing sequence is $(1/3!)(1 - x)^3\,dx$, because one out of the $3!$ permutations of the first three numbers has $x_1 > x_2 > x_3$ (required for a decreasing sequence).

Continuing in this manner, we see that the probability that a number between x and $x + dx$ is picked sooner or later in a decreasing sequence is

$$p(x)\,dx = \left(1 + (1 - x) + \frac{(1 - x)^2}{2!} + \frac{(1 - x)^3}{3!} + \cdots\right) dx$$
$$= e^{1-x}\,dx. \tag{3.180}$$

If we play a large number N of games, then we will have picked a total of $Ne^{1-x}\,dx$ numbers between x and $x + dx$ in the decreasing parts of all the sequences (that is, not counting the last number, which is larger than the previous one and which causes the game to end). The total number of numbers we pick in the decreasing parts of all the sequences is therefore $\int_0^1 Ne^{1-x}\,dx = N(e - 1)$. The average number of numbers per game in the decreasing part of the sequence is then $e - 1$. Adding on the last number which causes the game to end gives an average of e numbers per game.

REMARKS:

1. What is the average value of the smallest number you pick? The probability that the smallest number is between x and $x + dx$ equals $e^{1-x}(1 - x)\,dx$. This is true because from Eq. (3.180), $p(x)\,dx = e^{1-x}\,dx$ is the probability that you pick a number between x and $x + dx$ as part of the decreasing sequence, and then $(1 - x)$ is the probability that the next number you pick is larger. The average value, s, of the smallest number you pick is therefore $s = \int_0^1 e^{1-x}(1-x) \cdot x\,dx$. Letting $y \equiv 1 - x$ for convenience, and integrating by parts (or just looking up the integral), gives

$$s = \int_1^0 e^y y(1 - y)(-dy) = \int_0^1 e^y y(1 - y)\,dy$$
$$= \left(-y^2 e^y + 3y e^y - 3e^y\right)\Big|_0^1 = 3 - e \approx 0.282. \tag{3.181}$$

Likewise, the average value of the final number you pick is $\int_0^1 e^{1-x}(1-x) \cdot ((1+x)/2)\,dx$, which you can show equals $2 - e/2 \approx 0.64$. The $(1+x)/2$ in this integral arises from the fact that if you do pick a number greater than x, its average value is $(1+x)/2$.

2. We can also ask questions such as: Continue the game as long as $x_1 > x_2$, and $x_2 < x_3$, and $x_3 > x_4$, and $x_4 < x_5$, and so on, with the numbers alternating in size. What is the expected number of numbers you pick? The method of the second solution above works well here. (You should try to solve this before reading further.)

Let $A(x)$ be the expected number of numbers you have yet to pick after an odd pick (that is, for $x = x_1, x_3, x_5, \ldots$). At each of these stages, you are hoping that the next number is smaller. And let $B(x)$ be the expected number of numbers you have yet to pick after an even pick (that is, for $x = x_2, x_4, x_6, \ldots$). At each of these stages, you are hoping that the next number is larger. From the reasoning in the second solution, we have

$$A(x) = 1 \cdot (1 - x) + \int_0^x (B(y) + 1)\,dy = 1 + \int_0^x B(y)\,dy,$$

$$B(x) = 1 \cdot x + \int_x^1 (A(y) + 1)\,dy = 1 + \int_x^1 A(y)\,dy. \qquad (3.182)$$

Differentiating these two equations yields $A'(x) = B(x)$ and $B'(x) = -A(x)$. If we then differentiate the first of these relations and substitute the result into the second, we obtain $A''(x) = -A(x)$. (Likewise, $B''(x) = -B(x)$.) The solution to this equation is a $\sin x$ or $\cos x$ function, or more generally a linear combination, $A(x) = a \sin x + b \cos x$. $B(x)$ is then determined by $A'(x) = B(x)$, so we have

$$A(x) = a \sin x + b \cos x \qquad \text{and} \qquad B(x) = a \cos x - b \sin x. \qquad (3.183)$$

We can find the coefficients a and b by invoking two known values of A and B. First, we know that $A(0) = 1$, because if we have just picked 0 on an odd pick (after which we are hoping for a smaller number), then the next number will definitely be larger, in which case the game stops. This yields $b = 1$. Second, we know that $B(1) = 1$, because if we have just picked 1 on an even pick (after which we are hoping for a larger number), then the next number will definitely be smaller, in which case the game stops. This yields $a = (1 + \sin 1)/\cos 1$. (The angle "1" here is in radians.) The desired answer to the problem equals $B(0)$, because we could imagine starting the game with someone picking a number greater than 0, which is guaranteed. (Similarly, the desired answer also equals $A(1)$.) So the expected total number of picks is $B(0) = (1 + \sin 1)/\cos 1$. This has a value of about 3.41, which is greater than the $e \approx 2.72$ answer to our original problem. This makes intuitive sense; the monotonically decreasing sequence squeezes down the allowed range of future numbers more than the alternating sequence does. ♣

39. **Sum over 1**

(a) We will use the following fact: Given n random numbers between 0 and 1, the probability $P_n(1)$ that their sum does not exceed 1 equals $1/n!$. This quantity $1/n!$ is the volume of the n-dimensional region bounded by the coordinate planes and the hyperplane $x_1 + x_2 + \cdots + x_n = 1$. (For example, in two dimensions we have a triangle with area $1/2$, and in three dimensions we have a pyramid with volume $1/6$, etc.) This volume can be calculated in various ways, one of which is to evaluate the integral $\int_0^1 dx_1 \int_0^{1-x_1} dx_2 \int_0^{1-x_1-x_2} dx_3 \cdots$. You can work this out if you wish, but we'll demonstrate the $1/n!$ result by proving a slightly stronger theorem. (This theorem can alternatively be obtained via a slight tweak in the above multi-dimensional integral, as you can show.)

Theorem: *Given n random numbers between 0 and 1, the probability $P_n(s)$ that their sum does not exceed s equals $s^n/n!$, for all $s \leq 1$.*

Proof: Assume inductively that the result holds for a given n. (It certainly holds for all $s \leq 1$ when $n = 1$.) What is the probability that $n + 1$ numbers sum to no more than t (with $t \leq 1$)? Let the $(n+1)$th number have the value x. Then the probability $P_{n+1}(t)$ that all $n + 1$ numbers sum to no more than t equals the probability $P_n(t - x)$ that the first n numbers sum to no more than $t - x$, which is $P_n(t - x) = (t - x)^n/n!$ from the inductive hypothesis. (This hypothesis assumes that $t - x$ is less that 1. And since x can be as small as zero, we see that we must assume $t \leq 1$.) The probability that a number lies between x and $x + dx$ is just dx. So the probability that $n + 1$ numbers sum to no more than t, with the $(n+1)$th number lying between x and $x + dx$, is $dx \cdot (t - x)^n/n!$. Integrating this probability over all x from 0 to t gives

$$P_{n+1}(t) = \int_0^t \frac{(t-x)^n}{n!}\, dx = -\frac{(t-x)^{n+1}}{(n+1)!}\bigg|_0^t = \frac{t^{n+1}}{(n+1)!}. \quad (3.184)$$

We see that if the theorem holds for n, then it also holds for $n+1$. Therefore, since the theorem holds for all $s \leq 1$ when $n = 1$, it holds for all $s \leq 1$ for any n. ∎

We are concerned with the special case $s = 1$, in which case $P_n(1) = 1/n!$. The probability that it takes exactly n numbers for the sum to exceed 1 equals $1/(n-1)! - 1/n!$. This is true because the first $n - 1$ numbers must sum to less than 1, and the nth number must push the sum over 1, so we must subtract off the probability that it does not.

The expected number of numbers, N, to achieve a sum greater than 1, is therefore

$$N = \sum_2^\infty n\left(\frac{1}{(n-1)!} - \frac{1}{n!}\right) = \sum_2^\infty \frac{1}{(n-2)!} = e \approx 2.718, \quad (3.185)$$

which is as nice an answer as you could expect!

(b) Each of the random numbers has an average value of 1/2. Therefore, since it takes (on average) e numbers for the sum to exceed 1, the average value of the sum will be $e/2 \approx 1.36$.

This reasoning probably strikes you as being either completely obvious or completely mysterious. If the latter, imagine playing a large number of games in succession, writing down each of the random numbers in one long sequence. (You can note the end of each game by, say, putting a mark after the final number of that game, but this isn't necessary.) If you play N games (with N very large), then the result from part (a) tells us that there will be approximately Ne numbers listed in the sequence. Each number is a random number between 0 and 1, so the average value is 1/2. The sum of all the numbers in the sequence is therefore approximately $Ne/2$. Hence, the average sum per game is $e/2$.

40. **Convenient migraines**

(a) The student needs to have a headache (which occurs with probability p) on a specific days, and also to not have a headache (which occurs with probability $1 - p$) on b specific days. The probability of all of these events occurring is therefore

$$P(p) = p^a(1-p)^b. \tag{3.186}$$

Note that there is no need for a binomial coefficient here, because the exam days are fixed, so the desired series of events can happen in only one way. If $a = 2$ and $b = 18$, the plot of $P(p)$ is shown in Fig. 3.48 for p values up to 0.5, by which point $P(p)$ has become negligible.

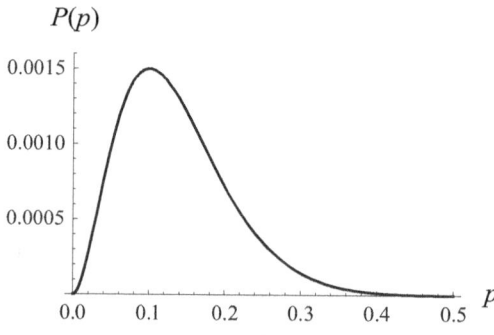

Figure 3.48

Maximizing the $P(p)$ in Eq. (3.186) by taking the derivative with respect to p gives

$$0 = \frac{dP}{dp} = ap^{a-1} \cdot (1-p)^b - p^a \cdot b(1-p)^{b-1}$$

$$\implies 0 = p^{a-1}(1-p)^{b-1}\Big(a(1-p) - bp\Big)$$

$$\implies p = \frac{a}{a+b}. \tag{3.187}$$

For $a = 2$ and $b = 18$, the value of p that maximizes $P(p)$ is therefore $p = 0.1$, which is consistent with a visual inspection of Fig. 3.48. The $p = a/(a + b)$ result checks in two limits: If $a \ll b$, then $p \approx 0$ (most days don't involve a migraine), and if $b \ll a$, then $p \approx 1$ (most days do involve a migraine).

(b) Substituting $p = a/(a + b)$ into Eq. (3.186) yields

$$P_{\max} = \left(\frac{a}{a+b}\right)^a \left(\frac{b}{a+b}\right)^b = \frac{a^a b^b}{(a+b)^{a+b}}. \tag{3.188}$$

For $a = 2$ and $b = 18$, this yields $P_{\max} \approx 1.50 \cdot 10^{-3} = 0.15\%$, consistent with Fig. 3.48. This is quite small, so it's fairly safe to say that the student was doing something shady.

Actually, this conclusion is a reasonable one *if* we accept the given assumption that the probability p of a migraine occurring on a given day is completely random and not based on real-life effects such as stress, etc. The more complete and correct conclusion is that either (a) it's likely that something shady was going on, or (b) our initial assumption was incorrect. In real life (which isn't an idealized math problem – sometimes for better, sometimes for worse), we need to somehow determine how good our assumptions are. In the present case, we can look at how stress affects the other students, although that still doesn't say anything definite about the student in question; maybe that student is simply more susceptible to stress. Getting data on headache occurrences in the given student's other courses might help, but maybe not much. Often, the most difficult part of a real-world problem is determining how reasonable the assumptions are. Life is complicated.

(c) We'll let $a \to z$, just to remind us that this is the quantity we're varying. $P_{\max}$ as a function of z is then

$$P_{\max}(z) = \frac{z^z b^b}{(z+b)^{z+b}} = \frac{z^z b^b}{b^{z+b}(z/b+1)^{z+b}} = \frac{z^z}{b^z} \frac{1}{(1+z/b)^{z+b}}. \tag{3.189}$$

Since we are assuming $z \ll b$, we can use the $(1+a)^n \approx e^{na}$ approximation from Problem 53 to write

$$\frac{1}{(1+z/b)^{z+b}} = \left(1+\frac{z}{b}\right)^{-(z+b)} \approx \left(1+\frac{z}{b}\right)^{-b} \approx e^{-z}. \tag{3.190}$$

(We have ignored the $(1+z/b)^{-z} \approx e^{-z^2/b}$ factor, which is essentially equal to 1 if z is sufficiently small, more precisely, if $z \ll \sqrt{b}$. Another factor involving $e^{-z^2/b}$ arises from the more accurate approximation in Eq. (1.6) in Problem 53 anyway, so it would be inconsistent to keep only the $e^{-z^2/b}$ term that arises here.) Eq. (3.189) then becomes

$$P_{\max}(z) \approx \frac{z^z}{b^z} \cdot e^{-z} = \left(\frac{z}{eb}\right)^z \to \left(\frac{a}{eb}\right)^a, \tag{3.191}$$

which is a nice clean result. For $a = 2$ and $b = 18$, this yields $P_{\max} \approx 1.67 \cdot 10^{-3}$, which is reasonably close to the actual $1.50 \cdot 10^{-3}$ result in part (b). Given a, the approximation gets better the larger b is. For example, if $a = 2$ and $b = 98$, the exact result for $P_{\max}$ in Eq. (3.188) is $5.52 \cdot 10^{-5}$, while the approximate result in Eq. (3.191) is $5.64 \cdot 10^{-5}$. And if $a = 2$ and $b = 998$, the exact result is $5.424 \cdot 10^{-7}$, while the approximate result is $5.435 \cdot 10^{-7}$.

41. **Letters in envelopes**

    FIRST SOLUTION: (Thanks to Aravi Samuel for this solution.) We will use induction on N. Let B_N denote the number of "bad" arrangements where none of the N letters end up in the correct envelope. We claim that

    $$B_{N+1} = N(B_N + B_{N-1}). \qquad (3.192)$$

    This can be seen as follows. In proceeding inductively from N to $N + 1$ letters, there are two possible ways we can generate bad arrangements:

    - Consider a bad arrangement of N letters. There are B_N of these, by definition. We can create a bad arrangement of $N + 1$ letters by transferring any one of the N letters to the $(N + 1)$th envelope (there are N ways to do this for each arrangement) and then filling the now-empty envelope with the $(N + 1)$th letter. This procedure provides us with NB_N possible bad arrangements of $N + 1$ letters.

    - Consider a arrangement of N letters having exactly one letter in the correct envelope. There are NB_{N-1} such arrangements, because for each choice of the correct letter, there are B_{N-1} bad arrangements of the other $N - 1$ letters. We can create a bad arrangement of $N + 1$ letters by transferring the correct letter to the $(N + 1)$th envelope and then filling the now-empty envelope with the $(N + 1)$th letter. This procedure provides us with NB_{N-1} possible bad arrangements of $N + 1$ letters.

    We therefore see that $B_{N+1} = N(B_N + B_{N-1})$. Since there are $N!$ possible arrangements involving N letters, the desired probability of obtaining a bad arrangement with N letters is $P_N = B_N/N!$. Hence $B_N = N! P_N$. So in terms of probabilities, the recursion relation in Eq. (3.192) becomes

    $$(N + 1)! P_{N+1} = N\Big(N! P_N + (N - 1)! P_{N-1}\Big)$$
    $$\implies (N + 1)P_{N+1} = NP_N + P_{N-1}. \qquad (3.193)$$

    To solve this recursion relation, we can write it in the more suggestive form,

    $$P_{N+1} - P_N = -\frac{1}{N + 1}\Big(P_N - P_{N-1}\Big). \qquad (3.194)$$

    Since $P_1 = 0$ and $P_2 = 1/2$, we have $P_2 - P_1 = 1/2$. We then find inductively that $P_k - P_{k-1} = (-1)^k/k!$. Therefore (writing 0 as $1 - 1$), our answer for P_N is

$$P_N = P_1 + \sum_{k=2}^{N}(P_k - P_{k-1})$$

$$= (1-1) + \sum_{k=2}^{N}\frac{(-1)^k}{k!}$$

$$= \sum_{k=0}^{N}\frac{(-1)^k}{k!}. \qquad (3.195)$$

This is the partial series expansion for e^{-1}. So for large N, P_N approaches $1/e \approx 37\%$. This series expansion for $1/e$ converges very rapidly, so N doesn't need to be very large for the approximation $P_N \approx 1/e$ to be a very good one. For example, if $N = 5$ we have $|P_5 - 1/e| \approx 0.001$.

REMARK: This $1/e$ result in the large-N limit can also be seen in the following (hand-wavy) way. The probability that a given letter does not end up in its corresponding envelope is $1 - 1/N$. Therefore, if we ignore the fact that the placements of the letters affect each other (because two letters cannot end up in the same envelope), then the probability that no letter ends up in the correct envelope is

$$\left(1 - \frac{1}{N}\right)^N \approx \frac{1}{e}. \qquad (3.196)$$

It isn't immediately obvious that the correlations between the letters can be neglected here, but in view of the above result, this is must be the case. ♣

SECOND SOLUTION: As above, let P_N be the probability that none of the N letters end up in the correct envelope. Let L_i and E_i denote the ith letter and corresponding ith envelope.

When the N letters are randomly put into N envelopes, consider a particular letter L_{a_1}. This letter L_{a_1} will end up in some envelope E_{a_2}. L_{a_2} will then end up in some E_{a_3}. L_{a_3} will then end up in some E_{a_4}, and so on. Eventually, one of the envelopes in this chain must be E_{a_1}. Let it be $E_{a_{n+1}}$. We may describe this situation by saying that L_{a_1} belongs to a "loop" of length n. If L_{a_1} ends up in its own envelope, then $n = 1$. If no letter ends up in the correct envelope, then the n's of the various loops can take on any values from 2 to N.

Claim: *When N letters are randomly put into N envelopes, the probability that the loop containing any particular letter L_{a_1} has length n is equal to $1/N$, independent of n.*

Proof: The claim is certainly true for $n = 1$, because any given letter has a $1/N$ probability of ending up in its own envelope. For a general value of $n > 1$, L_{a_1} has an $(N-1)/N$ probability of ending up in an E_{a_2} with $a_2 \neq a_1$. L_{a_2} then has an $(N-2)/(N-1)$ probability of ending up in an E_{a_3} with $a_3 \neq a_1$ (or a_2, since E_{a_2}

is already taken). This continues until $L_{a_{n-1}}$ has an $(N-(n-1))/(N-(n-2))$ probability of ending up in an E_{a_n} with $a_n \neq a_1$ (or $a_2, a_3, \ldots, a_{n-1}$, since those envelopes are already taken). Finally, L_{a_n} has a $1/(N-(n-1))$ probability of ending up in $E_{a_{n+1}} = E_{a_1}$. The probability that L_{a_1} belongs to a loop of length n is therefore equal to

$$\left(\frac{N-1}{N}\right)\left(\frac{N-2}{N-1}\right)\cdots\left(\frac{N-(n-1)}{N-(n-2)}\right)\left(\frac{1}{N-(n-1)}\right) = \frac{1}{N}. \blacksquare \quad (3.197)$$

Assume that a particular letter L_{a_1} belongs to a loop of length n (which happens with probability $1/N$). Then the probability that all of the $N-n$ other letters end up in the wrong envelopes is simply P_{N-n}, by definition. The probability that none of the N letters end up in the correct envelope is therefore

$$P_N = \frac{1}{N}\left(P_{N-2} + P_{N-3} + \cdots + P_1 + P_0\right). \quad (3.198)$$

The P_{N-1} term is missing from this relation, because a loop of length 1 would mean that L_{a_1} went into E_{a_1}. Note that $P_1 = 0$, and also $P_0 \equiv 1$ here.

Multiplying Eq. (3.198) through by N, and then subtracting the analogous equation for P_{N-1} (after multiplying through by $N-1$), gives

$$NP_N - (N-1)P_{N-1} = P_{N-2}$$
$$\implies P_N - P_{N-1} = -\frac{1}{N}(P_{N-1} - P_{N-2}). \quad (3.199)$$

This is the same as Eq. (3.194), with $N+1$ replaced by N. The solution proceeds as above.

THIRD SOLUTION: We will find P_N by explicitly counting the number, B_N, of "bad" arrangements where none of the N letters end up in the correct envelope, and then dividing B_N by the total number of possible arrangements, $N!$.

We can count the number, B_N, of bad arrangements in the following manner. There are $N!$ total arrangements. To count the number that have no letter in the correct envelope, we must subtract from $N!$ the number of arrangements with at least one letter in the correct envelope. So, for example, we must subtract the number of arrangements with (at least) L_1 in the correct envelope. There are $(N-1)!$ of these arrangements, because there are $(N-1)!$ permutations of the other $N-1$ letters. Likewise for the arrangements where another given one of the N letters is in the correct envelope. So there seem to be $N! - N(N-1)!$ arrangements with no letter in the correct envelope.

However, this result is incorrect (it equals zero, in fact), because we have double-counted some of the arrangements. For example, an arrangement that has (at least) L_1 and L_2 in the correct envelopes has been subtracted twice, whereas it should have been subtracted only once. There are $(N-2)!$ such arrangements, because there are $(N-2)!$ permutations of the other $N-2$ letters. Likewise for any of the $\binom{N}{2}$ pairs of letters. So we must add on $\binom{N}{2}(N-2)!$ arrangements.

But now an arrangement that has (at least) L_1, L_2, and L_3 in the correct envelopes has not been subtracted off at all. (This is true because we have subtracted it off $\binom{3}{1} = 3$ times since a triplet contains three individual letters. And then we have added it on $\binom{3}{2} = 3$ times since a triplet contains three pairs of letters). There are $(N-3)!$ such arrangements. Likewise for the other triplets ($\binom{N}{3}$ in all). We want to subtract each of them once, so we must subtract off $\binom{N}{3}(N-3)!$ arrangements.

Now, however, an arrangement that has (at least) L_1, L_2, L_3, and L_4 in the correct envelopes has been counted $-\binom{4}{1} + \binom{4}{2} - \binom{4}{3} = -2$ times (that is, we have subtracted it off twice). There are $(N-4)!$ such arrangements. Likewise for the other quadruplets ($\binom{N}{4}$ in all). We want to subtract each of them only once, so we must add on $\binom{N}{4}(N-4)!$ arrangements.

In general, if we have done this procedure up to $(k-1)$-tuples, then a given arrangement having (at least) k letters in the correct envelopes has been counted T times, where

$$T = -\binom{k}{1} + \binom{k}{2} - \cdots + (-1)^{k-1}\binom{k}{k-1}. \tag{3.200}$$

We now note that the binomial expansion gives

$$\begin{aligned} 0 &= (1-1)^k \\ &= 1 - \binom{k}{1} + \binom{k}{2} + \cdots + (-1)^{k-1}\binom{k}{k-1} + (-1)^k \\ &= 1 + T + (-1)^k. \end{aligned} \tag{3.201}$$

Therefore, $T = -2$ for even k, and $T = 0$ for odd k. (This is consistent with the results for the small values of k we dealt with above.) For every arrangement with at least one letter in the correct envelope, we want $T = -1$, because we want to subtract off the arrangement once from the total number $N!$ of possible arrangements. So for any given arrangement with (at least) k particular letters in the correct envelopes, Eq. (3.201) tells us that we have either undercounted it by one (for even k), or overcounted it by one (for odd k). (This is known as the *inclusion–exclusion principle*.) There are $(N-k)!$ such arrangements. Likewise for the other k-tuples ($\binom{N}{k}$ in all). We must therefore add on $(-1)^k\binom{N}{k}(N-k)!$ arrangements. The prefactor here takes the simple form of $(-1)^k$ because of our "over/undercounting by one" result. Hence the total number, B_N, of arrangements with no letter in the correct envelope is

$$\begin{aligned} N! - \binom{N}{1}(N-1)! + \binom{N}{2}(N-2)! + \cdots &= \sum_{k=0}^{N}(-1)^k \frac{N!}{k!(N-k)!} \cdot (N-k)! \\ &= \sum_{k=0}^{N} \frac{(-1)^k N!}{k!}. \end{aligned} \tag{3.202}$$

To obtain the probability, P_N, that no letter is in the correct envelope, we must divide this result by $N!$. Therefore,

$$P_N = \sum_{k=0}^{N} \frac{(-1)^k}{k!}. \tag{3.203}$$

REMARKS:

1. What is the probability (call it P_N^l) that exactly l out of the N letters end up in the correct envelopes? (With this notation, P_N^0 equals the P_N from above.) We can find P_N^l as follows. (You should think about this before reading further.)

   The probability that a given set of l letters all go into the correct envelopes is $1/\bigl(N(N-1)(N-2)\cdots(N-l+1)\bigr)$. (There is a $1/N$ chance for the first letter, then a $1/(N-1)$ chance for the second letter, etc.) The probability that the remaining $N-l$ letters all go into the wrong envelopes is P_{N-l}^0. This situation can happen in $\binom{N}{l}$ ways. Therefore,

$$P_N^l = \binom{N}{l} \cdot \frac{1}{N(N-1)\cdots(N-l+1)} \cdot P_{N-l}^0$$
$$= \frac{1}{l!} P_{N-l}. \tag{3.204}$$

   Hence, using Eq. (3.203)

$$P_N^l = \frac{1}{l!} \sum_{k=0}^{N-l} \frac{(-1)^k}{k!}. \tag{3.205}$$

   For a given l, if N is large then the above sum (excluding the $1/l!$ factor out front) is essentially equal to $1/e$, so we have $P_N^l \approx 1/(l!e)$. It then quickly follows that the sum of all the P_N^l probabilities (for l from 0 to N, assuming N is large) equals 1, as it must.

   The fact that $P_N^l \approx 1/(l!e)$ falls off so rapidly with l means that we are essentially guaranteed of having at most only a few letters in the correct envelopes. For example, we find (for large N) that the probability of having four or fewer letters in the correct envelopes is about 99.6%. Note that for large N, we have $P_N^0 \approx P_N^1$, with the common value being $1/e$.

2. It is interesting to note that the relation $P_N^l = (1/l!)P_{N-l}$ in Eq. (3.204) directly yields the large-N result, $P_N \approx 1/e$, without having to go through all the work of the original problem. To see this, we'll use the fact that the sum of all the P_N^l probabilities must be 1:

$$1 = \sum_{l=0}^{N} P_N^l = \sum_{l=0}^{N} \frac{1}{l!} P_{N-l}. \tag{3.206}$$

Since the terms with small l values dominate this sum, we may (for large N) replace the P_{N-l} values with $\lim_{M\to\infty} P_M$. Hence,

$$1 \approx \sum_{l=0}^{N} \frac{1}{l!} \left(\lim_{M\to\infty} P_M \right). \tag{3.207}$$

Therefore,

$$\lim_{M\to\infty} P_M \approx \left(\sum_{l=0}^{N} \frac{1}{l!} \right)^{-1} \approx \frac{1}{e}. \tag{3.208}$$

3. Let's check that the P_N^l probabilities Eq. (3.205) properly add up to 1 (for l from 0 to N), for any general value of N (not just large N). This can be done as follows. The sum we want to calculate is

$$\sum_{l=0}^{N} P_N^l = \sum_{l=0}^{N} \sum_{k=0}^{N-l} \frac{1}{l!} \frac{(-1)^k}{k!}. \tag{3.209}$$

(Try to show that this does indeed equal 1, before reading further.)

The range of l and k values forms a triangle in the l-k plane, as shown in Fig. 3.49 (we've chosen $N = 4$ for concreteness). We can get a handle on the sum by grouping the dots according to the dashed diagonal lines shown in the figure.

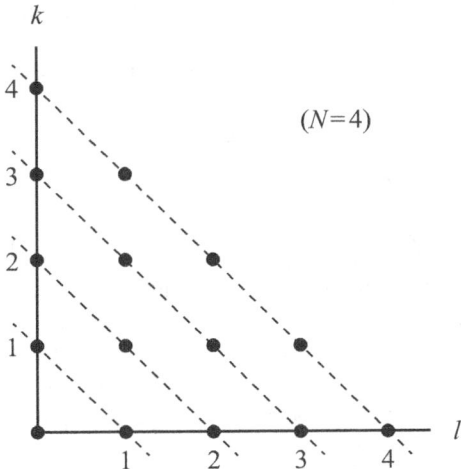

Figure 3.49

If we let $s \equiv l + k$, then the dashed lines are lines of constant s, where s runs from 0 to N. And for a given s value, l runs from 0 to s. So we can rewrite the sum as (with $k = s - l$, and using the binomial expansion to obtain the fourth line)

$$\sum_{l=0}^{N} P_N^l = \sum_{s=0}^{N} \sum_{l=0}^{s} \frac{1}{l!} \frac{(-1)^{s-l}}{(s-l)!}$$

$$= \sum_{s=0}^{N} \frac{1}{s!} \sum_{l=0}^{s} \frac{s!}{l!(s-l)!} (-1)^{s-l}$$

$$= \sum_{s=0}^{N} \frac{1}{s!} \sum_{l=0}^{s} \binom{s}{l} (-1)^{s-l}$$

$$= \sum_{s=0}^{N} \frac{1}{s!} (1-1)^s$$

$$= 1, \tag{3.210}$$

because only the $s = 0$ term contributes. In short, the sum along every dashed diagonal line in Fig. 3.49 is zero. Only the dot at the origin contributes.

4. What is the expected number, A, of letters in the correct envelopes? (Think about this before reading further.)

If the setup in the problem is repeated many times, then the expected number of times a given letter ends up in the correct envelope is $1/N$. Since there are N letters, the expected total number of correct envelopes is therefore $N(1/N) = 1$.

You can check that the expression for P_N^l in Eq. (3.205) leads to $A = \sum_{l=0}^{N} l \cdot P_N^l = 1$. For finite N, the technique used in Eq. (3.210) is helpful in evaluating this sum, if you want to work it out. For large N, where we have $P_N^l \approx 1/(l!e)$, the sum is easy, and we obtain

$$A = \sum_{l=0}^{N} l \cdot P_N^l \approx \frac{1}{e} \sum_{l=1}^{N} \frac{1}{(l-1)!} \approx \frac{1}{e} \cdot e = 1. \quad \clubsuit \tag{3.211}$$

42. **Leftover dental floss**

Let (x, y) denote the situation where x segments (of length d) have been cut off the right roll, and y segments have been cut off the left roll.[18] In solving this problem, we'll need to calculate the probability that the process ends at (N, n), in which case a length $(N - n)d$ remains on the left roll. (Or it can end at (n, N), in which case a length $(N - n)d$ remains on the right roll.) For this to happen, the process must first get to $(N - 1, n)$, and then the right roll must be chosen for the last segment. (The other way to get to (N, n), via the point $(N, n - 1)$,

---

[18] You might think that this convention is backwards, in that the left coordinate of the ordered pair should correspond to the left roll, and the right coordinate should correspond to the right roll. This is a perfectly reasonable convention, but we chose the opposite one because in visualizing the process, it helps to map it onto the two-dimensional x-y plane (see the remark at the end of the solution); and the left coordinate in an ordered pair corresponds to *rightward* motion in the plane. In any case, there's no right or wrong convention; it's just personal preference.

doesn't apply here, because in that case the process would have already ended at $(N, n-1)$.) The probability of reaching the point $(N-1, n)$ is

$$P_{N-1,n} = \frac{1}{2^{N-1+n}} \binom{N-1+n}{n}, \tag{3.212}$$

because the binomial coefficient gives the number of different ways the left roll can be chosen n times during the total number of $N-1+n$ choices (each of which involves a probability of $1/2$). The probability of then choosing the right roll for the next piece is $1/2$. Therefore, the probability of ending the process at (N, n) is

$$P_{N,n}^{\text{end}} = \frac{1}{2^{N+n}} \binom{N-1+n}{n}. \tag{3.213}$$

By the same reasoning, this is also the probability of ending the process at (n, N). In each of these cases the leftover length is $(N-n)d$. So if we take into account all the possible values of n (from 0 to $N-1$), we see that the average leftover length at the end of the process is

$$\ell = \sum_{n=0}^{N-1} (N-n)d \cdot P_{N,n}^{\text{end}} + \sum_{n=0}^{N-1} (N-n)d \cdot P_{n,N}^{\text{end}}. \tag{3.214}$$

The two sums here are the same, so we can deal with just the first one and multiply by 2. Using Eq. (3.213), we obtain

$$\ell = 2d \sum_{n=0}^{N-1} \frac{N-n}{2^{N+n}} \binom{N-1+n}{n}. \tag{3.215}$$

This is the exact answer to the problem, but it isn't very enlightening. So let's generate an approximate form of the answer (valid for large N) that makes it far easier to see the dependence on N. If you numerically perform the above sum for a few large values of N, it becomes clear that ℓ grows like $\sqrt{N}$. Let's show this analytically.

In order to make an approximation to Eq. (3.215), we will use the standard fact that for large N, a binomial coefficient can be approximated by a Gaussian function. From Eqs. (1.10) and (1.11) in Problem 56 we have, for large M and $x \ll M$,

$$\binom{2M}{M-x} \approx \frac{2^{2M}}{\sqrt{\pi M}} e^{-x^2/M}. \tag{3.216}$$

To make use of this, we'll first need to rewrite Eq. (3.215) as (with $z \equiv N - n \Longrightarrow n = N - z$)

$$\ell = 2d \sum_{n=0}^{N-1} \frac{N-n}{2^{N+n}} \frac{N}{N+n} \binom{N+n}{n}$$

$$= 2d \sum_{z=1}^{N} \frac{z}{2^{2N-z}} \frac{N}{2N-z} \binom{2N-z}{N-z}$$

$$= 2d \sum_{z=1}^{N} \frac{z}{2^{2N-z}} \frac{N}{2N-z} \binom{2(N-z/2)}{(N-z/2)-z/2}. \tag{3.217}$$

With $M \equiv N - z/2$ and $x \equiv z/2$, using Eq. (3.216) to rewrite the binomial coefficient gives our desired approximate answer as (see below for an explanation of the steps)

$$\begin{aligned} \ell &\approx 2d \sum_{z=1}^{N} \frac{Nz}{2N-z} \frac{e^{-z^2/4(N-z/2)}}{\sqrt{\pi(N-z/2)}} \\ &\approx \frac{d}{\sqrt{\pi N}} \sum_{z=1}^{N} z e^{-z^2/(4N)} \\ &\approx \frac{d}{\sqrt{\pi N}} \int_{0}^{\infty} z e^{-z^2/(4N)} \, dz \\ &= -\frac{d}{\sqrt{\pi N}} \cdot 2N e^{-z^2/(4N)} \bigg|_{0}^{\infty} \\ &= 2d\sqrt{\frac{N}{\pi}} \, . \end{aligned} \tag{3.218}$$

In obtaining the second line above, we have kept only the terms of leading order in N. The exponential factor guarantees that only z values up to order $\sqrt{N}$ will contribute. Hence, z is negligible when added to N. In obtaining the third line, we have used the fact that since N is large, the sum can be approximated by an integral; the (integer) values of z are effectively continuous. And the integral can be extended to infinity with negligible error, because large values of z (much larger than $\sqrt{N}$) contribute negligibly. Likewise, the lower limit can be dropped to zero, because the error introduced (which is less than 1) is much less than the value of the integral itself (which is of order N).

In terms of the initial length of floss in each roll, $L \equiv Nd$, the average leftover length in Eq. (3.218) can be written as $\ell \approx (2/\sqrt{\pi})\sqrt{Ld}$, which is proportional to the geometric mean of L and d. So increasing L by a factor of, say, 10 and decreasing d by the same factor of 10 (in which case N increases by 100) will lead to the same average leftover length.

REMARK: Geometrically, this problem is equivalent to the following one. Draw a square with one corner at the origin and the opposite corner at the point $(L, L) = (Nd, Nd)$. Start at the origin and take steps of length d in the x or y direction, with equal probabilities. (A rightward step corresponds to taking a piece of floss from the right roll, and an upward step corresponds to taking a piece from the left roll.) When you reach the $x = Nd$ or $y = Nd$ side of the box, how far are you from the corner at (Nd, Nd)? This distance corresponds to the amount of leftover dental floss in the original problem. A possible path is shown in Fig. 3.50, where we have chosen $N = 6$ for concreteness. Without doing any calculations, it's a good bet that in a random-walk problem like this, the answer should go like $\sqrt{Nd}$ for large N. But it takes some effort to show, as we did above, that the coefficient is $2/\sqrt{\pi}$. ♣

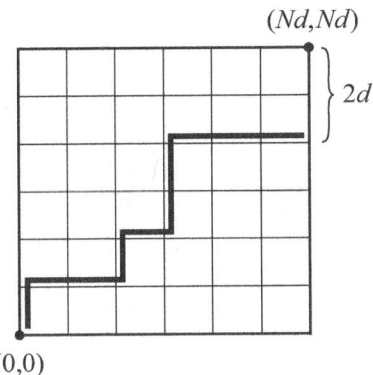

Figure 3.50

43. **Comparing the numbers**

   (a) Let your number be n. We will average over the equally likely values of n (excluding $n = 1$) at the end of the calculation.

   For convenience, let $p_n \equiv (n-1)/(N-1)$ be the probability that a person you ask has a number smaller than yours. Then $1 - p_n \equiv (N-n)/(N-1)$ is the probability that a person you ask has a number larger than yours. Remember that p_n is constant for all of the steps here in part (a), because you are assumed to have a bad memory.

   Let A_n be the average number of people you need to ask in order to find a number smaller than yours, given that you have the number n. A_n may be calculated as follows. We'll give three derivations.

   FIRST DERIVATION: There is a probability p_n that it takes only one check to find a smaller number.

   There is a probability $1 - p_n$ that the first person you ask has a larger number. From this point on, you need to ask (by definition) an average of A_n people in order to find a smaller number. (This is true because you could imagine starting the game at that point. Any starting point is as good as any other; A_n doesn't depend on which one, since you have no memory.) In this scenario, you end up asking a total of $A_n + 1$ people.

   Putting the above two possibilities together, we see that

   $$A_n = p_n \cdot 1 + (1 - p_n) \cdot (A_n + 1). \tag{3.219}$$

   Solving for A_n yields

   $$A_n = \frac{1}{p_n} = \frac{N-1}{n-1}. \tag{3.220}$$

   For example, if $n = N$, then you (always) need to ask only one person, because every other number is smaller than yours. And if $n = 2$, then you (on average) need to ask $N - 1$ people. This makes sense; if you check a million numbers in a row, about $1/(N-1)$ of them will be 1's. So on average you will need to check $N-1$ numbers, from one 1 to the next (that is, from the end of one game to the end of the next game). The third derivation below uses this line of reasoning.

SECOND DERIVATION: There is a probability p that it takes only one check to find a smaller number. (We'll drop the subscript n from p_n in this derivation, to keep the equations from getting too cluttered.)

There is a probability $(1-p)p$ that it takes two checks to find a smaller number, because in this case the first person you ask must have a larger number (with probability $1-p$), and then the second person you ask must have a smaller number (with probability p).

There is a probability $(1-p)^2 p$ that it takes three checks to find a smaller number, because in this case the first two people you ask must have larger numbers (each with probability $(1-p)$), and then the third person you ask must have a smaller number (with probability p).

This pattern continues, so we see that the expected number of checks is

$$A_n = 1 \cdot p + 2 \cdot (1-p)p + 3 \cdot (1-p)^2 p + 4 \cdot (1-p)^3 p + \cdots . \quad (3.221)$$

We can evaluate this sum by recognizing that it can be written as

$$\begin{aligned} & p + (1-p)p + (1-p)^2 p + (1-p)^3 p + \cdots \\ & \quad (1-p)p + (1-p)^2 p + (1-p)^3 p + \cdots \\ & \quad\quad\quad\quad (1-p)^2 p + (1-p)^3 p + \cdots \\ & \quad\quad\quad\quad\quad\quad\quad\quad (1-p)^3 p + \cdots \\ & \quad\quad\quad\quad\quad\quad\quad\quad\quad\quad \vdots \end{aligned} \quad (3.222)$$

This has the correct number of each type of term. For example, $(1-p)^2 p$ appears three times. The first line is an infinite geometric series that sums to $a_0/(1-r) = p/\bigl(1-(1-p)\bigr) = 1$. The second line is also an infinite geometric series, and it sums to $(1-p)p/\bigl(1-(1-p)\bigr) = 1-p$. Likewise the third line sums to $(1-p)^2$. And so on. The sum of the infinite number of lines in Eq. (3.222) therefore equals

$$1 + (1-p) + (1-p)^2 + (1-p)^3 + \cdots . \quad (3.223)$$

But this itself is an infinite geometric series, and it sums to $a_0/(1-r) = 1/\bigl(1-(1-p)\bigr) = 1/p$. So $A_n = 1/p \equiv 1/p_n$, in agreement with Eq. (3.220).

THIRD DERIVATION: Imagine playing many games in succession (all with the same particular number n for you), and writing down a long string of L's (if you find a number lower than yours) or H's (if you find a number higher than yours). Each game ends when you get an L. For example, in the (somewhat short) string LHLLHHHHLHHL, there are five games with lengths 1, 2, 1, 5, and 3. If the string contains ℓ (which is assumed to be large) letters, then the number of L's that appear is (approximately) $p_n \ell$, because each L occurs with probability p_n. Therefore, since $p_n \ell$ L's appear in the string with length ℓ, the average number of letters from one L to the next (which is the average length of a game) is $\ell/(p_n\ell) = 1/p_n$.

Having found A_n, we now note that since all values of n (from 2 to N) are equally likely, the desired average is simply the average of the numbers A_n, for n ranging from 2 to N. The average of these $N - 1$ numbers is

$$A = \frac{1}{N-1} \sum_{n=2}^{N} \frac{N-1}{n-1}$$
$$= 1 + \frac{1}{2} + \frac{1}{3} + \cdots + \frac{1}{N-1}. \tag{3.224}$$

This expression for A is the exact answer to the problem. To obtain an approximate answer for large N, we can invoke the fact that the sum of the reciprocals of the numbers from 1 to M approaches (for large M) $\ln M + \gamma$, where $\gamma \approx 0.577$ is Euler's constant. So if N is large, you need to check about $\ln(N-1) + \gamma \approx \ln N + \gamma$ other numbers before you find one that is smaller than yours.

(b) Let your number be n. As in part (a), we will average over the equally likely values of n (excluding $n = 1$) at the end of the calculation.

Let B_n^N be the average number of people you need to ask in order to find a number smaller than yours, given that you have the number n. (We've added the index (not exponent!) N here, which wasn't present in the A_n notation in part (a), because our strategy below will be to produce a recursion relation in N.) B_n^N may be calculated as follows.

There is a probability $(n-1)/(N-1)$ that it takes only one check to find a smaller number.

There is a probability $(N-n)/(N-1)$ that the first person you ask has a larger number. From this point on, you need to ask (by definition) an average of B_n^{N-1} people in order to find a smaller number, because you won't ask the first person again. (The indices on B are $N-1$ and n, because it doesn't matter which of the numbers larger than yours you encountered, and because there are still the same $n-1$ numbers smaller than yours out there.) In this scenario, you end up asking a total of $B_n^{N-1} + 1$ people.

Putting the above two possibilities together, we see that

$$B_n^N = \frac{n-1}{N-1} \cdot 1 + \frac{N-n}{N-1} \cdot \left(B_n^{N-1} + 1\right)$$
$$= 1 + \left(\frac{N-n}{N-1}\right) B_n^{N-1}. \tag{3.225}$$

Using the fact that $B_n^N = 1$ when $N = n$ (assuming that $n \neq 1$), we can use the recursion relation in Eq. (3.225) to inductively increase N (while holding n constant) to obtain B_n^N for $N > n$. If you work out a few cases, you will quickly see that $B_n^N = N/n$. We can then easily check this by induction on N; it is true for $N = n$, so we simply need to verify in Eq. (3.225) that

$$\frac{N}{n} = 1 + \left(\frac{N-n}{N-1}\right)\frac{N-1}{n}, \tag{3.226}$$

which is indeed true. Therefore,

$$B_n^N = \frac{N}{n}. \tag{3.227}$$

This result isn't valid when $n = 1$, because the starting point in the induction, namely $B_n^N = 1$ when $N = n$, isn't valid when $n = 1$. B_1^1 actually isn't even defined, since there are no other numbers available to check. B_1^N isn't defined for larger N either, because you will always end up checking all of the other $N - 1$ numbers and never being successful in finding a smaller one.

As in part (a), all values of n (from 2 to N) are equally likely, so we simply need to find the average of the numbers $B_n^N = N/n$, for n ranging from 2 to N. The average of these $N - 1$ numbers is

$$B = \frac{1}{N-1} \sum_{n=2}^{N} \frac{N}{n}$$

$$= \frac{N}{N-1}\left(\frac{1}{2} + \frac{1}{3} + \cdots + \frac{1}{N}\right). \tag{3.228}$$

This expression for B is the exact answer to the problem. If N is large, then the result is approximately equal to $\ln N + \gamma - 1$, due to the first term of "1" missing in the parentheses. This result is 1 less than the result in part (a). So for large N, a good memory saves you, on average, one query. (Most people would probably guess that it saves more than that.) Note that $n = 2$ yields the largest difference between the B_n^N in Eq. (3.227) and the A_n in Eq. (3.220) ($N/2$ vs. $N - 1$). That is, your memory matters the most when $n = 2$. For small n, you're probably going to be checking a lot of numbers that are larger than yours, so it helps if you can avoid checking them more than once.

EXTENSION: The continuum version of this problem is the following. (Note that the quality of your memory is irrelevant now, assuming that you pick random numbers to, say, 20 decimal places. Even with a bad memory, there is virtually no chance that you pick the same number twice.)

Someone gives you a random number between 0 and 1, with a flat distribution. Pick successive random numbers between 0 and 1 until you obtain one that is smaller. How many numbers, on average, will you need to pick? (You should think about this before reading further.)

This is simply the original problem, in the $N \to \infty$ limit. So up to the additive constant γ, the answer should be $\approx \ln N \to \ln(\infty)$, which is infinite. And indeed, from the reasoning in part (a), if you start with the number x, the average number of picks you need to make to find a smaller number is

$1/p = 1/x$, from Eq. (3.220). Averaging these waiting times of $1/x$, over the equally likely values of x, gives an average waiting time of

$$\int_0^1 \frac{dx}{x} = \infty. \tag{3.229}$$

However, if you play this game a few times, you will quickly discover that your average number of necessary picks is not infinite. If you find this unsettling, you are encouraged to look at Problem 25 (Flipping a coin).

44. Shifted intervals

Let us discretize each of the intervals (of length 1) into units of length ϵ (which is very small, since N is very large). If the first number is in the smallest of its possible ϵ-units (that is, between 0 and ϵ), then it is guaranteed to be the smallest of all the numbers. If the first number is in the second smallest ϵ-unit (between ϵ and 2ϵ), then there is a $1 - \epsilon$ chance that it is the smallest of all the numbers, because this is the probability that the second number is larger than it.[19] (All the other numbers are guaranteed to be larger.) If the first number is in the third ϵ-unit (between 2ϵ and 3ϵ), then there is a $(1 - \epsilon)(1 - 2\epsilon)$ chance that it is the smallest, because this is the probability that both the third and second numbers are larger than it. (All the other numbers are guaranteed to be larger.) In general, if the first number is in the kth ϵ-unit, then there is a

$$P_k = (1 - \epsilon)(1 - 2\epsilon)(1 - 3\epsilon) \cdots (1 - (k-1)\epsilon) \tag{3.230}$$

chance that it is the smallest. Since the first number has an equal probability of ϵ of being in any of the ϵ-units, the total probability that it is the smallest number is

$$P = \epsilon P_1 + \epsilon P_2 + \epsilon P_3 + \cdots + \epsilon P_N. \tag{3.231}$$

For small ϵ, we can make an approximation to the P_k's, as follows. Take the log of P_k in Eq. (3.230) to obtain

$$\ln P_k = \ln(1 - \epsilon) + \ln(1 - 2\epsilon) + \ln(1 - 3\epsilon) + \cdots + \ln(1 - (k-1)\epsilon), \tag{3.232}$$

and then apply the Taylor series $\ln(1 - x) \approx -x - x^2/2$. (See the appendix for a review of Taylor series.) We'll see below that we don't actually need to include the $x^2/2$ term here, but we'll keep it to be safe. We then have

$$\ln P_k \approx \left(-\epsilon - \frac{\epsilon^2}{2}\right) + \left(-2\epsilon - \frac{2^2\epsilon^2}{2}\right) + \cdots + \left(-(k-1)\epsilon - \frac{(k-1)^2\epsilon^2}{2}\right)$$

$$= -\epsilon\Big(1 + 2 + \cdots + (k-1)\Big) - \frac{\epsilon^2}{2}\Big(1 + 2^2 + \cdots + (k-1)^2\Big)$$

$$= -\epsilon\left(\frac{k(k-1)}{2}\right) - \frac{\epsilon^2}{2}\left(\frac{k(k-1)(2k-1)}{6}\right)$$

$$\approx -\frac{\epsilon k^2}{2} - \frac{\epsilon^2 k^3}{6}. \tag{3.233}$$

---

[19]Technically, the probability is on average equal to $1 - \epsilon/2$, because the average value of the first number in this case is $3\epsilon/2$. However, the $\epsilon/2$ correction in this probability (and other analogous ones) is inconsequential, as we'll see.

We have used the fact that the k values we will be concerned with are generally large, which means that we need to keep only the leading powers of k. Exponentiating Eq. (3.233) and replacing ϵ with $1/N$ gives

$$P_k \approx e^{-k^2/2N} e^{-k^3/6N^2}. \tag{3.234}$$

The second factor here is essentially equal to 1 if $k^3/N^2 \ll 1$, that is, if $k \ll N^{2/3}$. But we are concerned only with k values up to order $N^{1/2}$, because if k is much larger than this, the first exponential factor in Eq. (3.234) makes P_k essentially zero. Since $N^{1/2} \ll N^{2/3}$ when N is large, we see that whenever P_k is not essentially zero, we can set the second exponential factor equal to 1. So we have[20]

$$P_k \approx e^{-k^2/2N}. \tag{3.235}$$

Eq. (3.231) then becomes (with $\epsilon = 1/N$)

$$P \approx \frac{1}{N} \left(e^{-1^2/2N} + e^{-2^2/2N} + e^{-3^2/2N} + \cdots + e^{-N^2/2N} \right). \tag{3.236}$$

Since N is large, successive terms here differ by only a small amount. So we can approximate the sum by an integral. And since the terms eventually become negligibly small, we can let the integral run to infinity, with negligible error. We then have

$$P \approx \frac{1}{N} \int_0^\infty e^{-z^2/2N} dz. \tag{3.237}$$

Using the general result, $\int_{-\infty}^\infty e^{-y^2/b} dy = \sqrt{\pi b}$ (see Eq. (3.280) in the solution to Problem 52 for a proof), we obtain

$$P \approx \frac{1}{2} \cdot \frac{1}{N} \sqrt{\pi \cdot 2N} = \sqrt{\frac{\pi}{2N}}. \tag{3.238}$$

For example, if $N = 10^4$, then $P \approx 1.3\%$. Note that since the P_k in Eq. (3.235) is negligibly small if $k \gg \sqrt{N}$, most of the terms in the sum in Eq. (3.236) are effectively zero. The fraction of the terms that contribute goes like $\sqrt{N}/N = 1/\sqrt{N}$. That is, if your first number is much larger than $1/\sqrt{N}$, there is a negligible chance that it ends up being the smallest. You can use this fact to show that the $\epsilon/2$ corrections we mentioned in Footnote 19 are indeed inconsequential. The exponent in Eq. (3.235) will pick up a term of order k/N (as you can show), and $e^{k/N}$ is essentially equal to 1 for the k values we're concerned with (ones that aren't much larger than $\sqrt{N}$). Equivalently, we're dropping a term of subleading order in k, just as we did in Eq. (3.233).

Eq. (3.238) shows that P scales like $1/\sqrt{N}$. So, for example, if $N = 10^6$, then $P \approx 0.13\%$, which is $\sqrt{100} = 10$ times smaller than the $P \approx 1.3\%$ result we found above for $N = 10^4$.

---

[20]If k is small, then we don't have any right to drop terms of subleading order in k in Eq. (3.233). However, if k is small, then $e^{-k^2/2N}$ is essentially equal to 1 (no matter what small k terms we put in the numerator of the exponent), because N is large. So, for example, using $P_1 \approx e^{-1^2/2N}$ in place of the true $P_1 = 1$ value produces negligible error.

REMARK: If we consider a different setup where all of the N intervals have the same range from 0 to 1, instead of being successively shifted by ϵ, then the probability that the first number is the smallest is simply $1/N$, because any one of the N numbers you pick is equally likely to be the smallest. (This result is exact, whereas the result in Eq. (3.238) is approximate.) It makes sense that the $\propto 1/\sqrt{N}$ probability for the shifted intervals is larger than the $1/N$ probability for the non-shifted intervals.

If you want to derive the "non-shifted" $1/N$ result by doing an integral (analogous to the discrete sum in Eq. (3.231)), observe that if the first number equals x, then there is a $(1-x)^{N-1}$ chance that all of the other $N-1$ numbers are larger than x. Therefore (using the fact that dx is the probability that the first number lies between x and $x + dx$),

$$P = \int_0^1 (1-x)^{N-1}\, dx = -\frac{(1-x)^N}{N}\bigg|_0^1 = \frac{1}{N}. \tag{3.239}$$

Alternatively, assuming that N is large, $(1-x)^{N-1}$ is non-negligible only for small x, in which case Eq. (1.5) from Problem 53 gives $(1-x)^{N-1} \approx e^{-(N-1)x} \approx e^{-Nx}$. So for large N,

$$P \approx \int_0^1 e^{-Nx}\, dx = -\frac{e^{-Nx}}{N}\bigg|_0^1 \approx \frac{1}{N}. \tag{3.240}$$

If the first number is much larger than $1/N$, the above e^{-Nx} result tells us that there is a negligible chance that this number ends up being the smallest. In the original problem with the shifted intervals, we found (see the paragraph following Eq. (3.238)) that the transition to negligible probability occurs at order $1/\sqrt{N}$. It makes sense that the order-$1/\sqrt{N}$ transition for the shifted intervals is larger than the order-$1/N$ transition for the non-shifted intervals. ♣

45. Intervals between independent events

(a) To find the average value (or expectation value) of a quantity, we must multiply each value by the probability of the value occurring, and then integrate over all the values. So, using the $pe^{-pt}\, dt$ result from Problem 54, the average waiting time (starting at any given time, not necessarily the time of an event) until the next event is

$$t_{\text{avg}} = \int_0^\infty t \cdot pe^{-pt}\, dt = -e^{-pt}\left(t + \frac{1}{p}\right)\bigg|_0^\infty = \frac{1}{p}, \tag{3.241}$$

as desired. (You can verify the integral here by differentiating it.) It makes sense that this average time decreases with p; if p is large, the events happen frequently, so the waiting time is short.

Since this $1/p$ result holds for any arbitrary starting time, we are free to choose the starting time to be the time of an event. A special case of this result is therefore the statement that the average waiting time between events is $t_{\text{between}} = 1/p$. This is consistent with the fact that $t/t_{\text{between}} = t/(1/p) = pt$ is the average number of events that occur during a (not necessarily infinitesimal) time t.

(b) If we pick a random point in time, then the average waiting time until the next event is $1/p$, from part (a). And the average time *since* the previous event is also $1/p$, because we can use the same reasoning that we used in part (a), going backward in time, to calculate the probability that the most recent event occurred at a time between t and $t + dt$ earlier. The direction of time is irrelevant; the process is completely described by saying that $p\,dt$ is the probability of an event happening in an infinitesimal time dt, and this makes no reference to a direction of time. The average length of the interval surrounding a randomly chosen point in time is therefore $1/p + 1/p = 2/p$.

(c) The $pe^{-pt}\,dt$ result from Problem 54 tells us (with the starting time of an interval chosen to be the time of an event) that an event-to-event interval with length between t and $t + dt$ occurs with probability $pe^{-pt}\,dt$. (That is, out of a billion successive intervals, roughly $(10^9)(pe^{-pt}\,dt)$ of them will have this length.) But if you pick a random point in time, $pe^{-pt}\,dt$ is *not* the probability that you will end up in an interval with length between t and $t + dt$, because *you are more likely to end up in an interval that is longer.*

Consider the simple case where there are only two possible lengths of intervals, 1 and 100, and these occur with equal probabilities of $1/2$. If you look at 1000 successive intervals, then about 500 will have length 1, and about 500 will have length 100. But if you pick a random point in time, you are 100 times more likely to end up in one of the large intervals. The probability of falling in each type of interval is *not* $1/2$. The two probabilities are instead $1/101$ and $100/101$. The probability ($1/101$ or $100/101$ here) of falling in an interval of a given length does *not* equal the probability ($1/2$ and $1/2$ here) of that given length occurring at a particular point in a list of all the lengths. In this example, the average distance between events is 50.5, while the average distance surrounding a randomly chosen point is, as you can show, 99.02. (These results don't have anything to do with the above results involving p, because the present example isn't a random process described by a given probability per unit time (or distance, or whatever). But it illustrates the basic point.)

In short, the probability of falling in an interval with length between t and $t + dt$ is proportional both to $pe^{-pt}\,dt$ (because the more intervals there are of a certain length, the more likely you are to land in one of them), *and* to the length t of the intervals (because the longer they are, the more likely you are to land in one of them).

(d) Consider a large number N of intervals. The *number* of intervals with length between t and $t+dt$ is $N(pe^{-pt}\,dt)$. The total *length* of these intervals (ones with length between t and $t + dt$) is therefore $N(pe^{-pt}\,dt) \cdot t$. The total length of *all* of the N intervals is the integral of this, which we quickly see (using Eq. (3.241)) equals N/p, as it should (because the average length of an interval is $1/p$).

The probability of picking a point in time that falls in one of the intervals with length between t and $t + dt$ equals the total length associated with these intervals, divided by the total length of all of the N intervals, which gives $(Npe^{-pt}t\,dt)/(N/p) = p^2 e^{-pt}t\,dt$. (Looking back at Eq. (3.241), we see that the integral of this probability equals 1, as it must.) As mentioned in part (c), this probability is proportional to both $pe^{-pt}\,dt$ and t. The expectation value of the length of the interval that the given point falls in is obtained by multiplying this probability by the interval length t, and then integrating. This gives

$$\int_0^\infty p^2 e^{-pt} t^2 \, dt = -\frac{e^{-pt}}{p}\left(2 + 2pt + p^2 t^2\right)\Big|_0^\infty = \frac{2}{p}, \qquad (3.242)$$

as desired. (Again, you can verify this integral by differentiating it.)

To sum up, there are two different probabilities in this problem: (1) the probability that a randomly chosen interval has length between t and $t + dt$ (this equals $pe^{-pt}\,dt$), and (2) the probability that a randomly chosen point in time falls in an interval with length between t and $t + dt$ (this equals $p^2 e^{-pt} t\, dt$). In the first case, by "randomly" we mean that we label each interval with a number and then pick a random number. The length of each interval is irrelevant in this case, whereas it is quite relevant in the second case.

46. **The prosecutor's fallacy**

We'll assume that we are concerned only with people living in Boston. There are one million such people, so if one person in 10,000 fits the description, this means that there are 100 people in Boston who fit it (one of whom is the perpetrator). When the police officers pick up someone fitting the description, this person could be any one of these 100 people. So the probability that the defendant in the courtroom is the actual perpetrator is only 1/100. In other words, there is a 99% chance that the person is innocent. A guilty verdict (based on the given evidence) would therefore be a horrible and tragic vote.

The above (correct) reasoning is fairly cut and dry, but it contradicts the prosecutor's reasoning. That reasoning must therefore be incorrect. But what exactly is wrong with it? It seems quite plausible at every stage. To isolate the flaw in the logic, let's list out the three separate statements the prosecutor made in his argument:

1. Only one person in 10,000 fits the description.
2. It is highly unlikely (far beyond a reasonable doubt) that an innocent person fits the description.
3. It is therefore highly unlikely that the defendant is innocent.

As we noted when we posed the problem, the first two of these statements are true. Statement 1 is true by assumption, and Statement 2 is true basically because 1/10,000 is a small number. Let's be precise about this and work out the exact probability that an innocent person fits the description. Of the one million people in Boston, the number who fit the description is $(1/10{,}000)(10^6) = 100$. Of these 100 people, only one is guilty, so 99 are innocent. And the total number of innocent people is $10^6 - 1 = 999{,}999$. The probability that an innocent person fits the description is therefore

$$\frac{\text{innocent and fits description}}{\text{innocent}} = \frac{99}{999{,}999} \approx 9.9 \cdot 10^{-5} \approx \frac{1}{10{,}000}. \quad (3.243)$$

As expected, the probability is essentially equal to 1/10,000.

Now let's look at the third statement above. This is where the error is. This statement is false, because Statement 2 simply does not imply Statement 3. We know this because we have already calculated the probability that the defendant is innocent, namely 99%. This correct probability of 99% is vastly different from the incorrect probability of 1/10,000 that the prosecutor is trying to mislead you with. However, even though the correct result of 99% tells us that Statement 3 must be false, where exactly is the error? After all, at first glance Statement 3 *seems* to follow from Statement 2. The error is the confusion of conditional probabilities. In detail:

- Statement 2 deals with the probability of fitting the description, *given* innocence. The (true) statement is equivalent to "*If* a person is innocent, *then* there is a very small probability that he fits the description." This probability is the conditional probability $P(D|I)$ (read as "the probability of D, given I"), with D for description and I for innocence.

- Statement 3 deals with the probability of innocence, *given* that the description is fit. The (false) statement is equivalent to "*If* a person (such as the defendant) fits the description, *then* there is a very small probability that he is innocent." This probability is the conditional probability $P(I|D)$.

These two conditional probabilities are *not* the same. The error is the assumption (or implication, on the prosecutor's part) that they are. As we saw above, $P(D|I) = 99/999{,}999 \approx 0.0001$, whereas $P(I|D) = 0.99$. These two probabilities are markedly different.

Intuitively, $P(D|I)$ is very small because a very small fraction of the population (in particular, a very small fraction of the innocent people) fit the description. And $P(I|D)$ is very close to 1 because nearly everyone (in particular, nearly everyone who fits the description) is innocent. This state of affairs is indicated in Fig. 3.51. (This a just a rough figure; the areas aren't actually in the proper proportions.) The large oval represents the 999,999 innocent people, and the small oval represents the 100 people who fit the description.

There are three basic types of people in the figure: There are $A = 999{,}900$ innocent people who don't fit the description, $B = 99$ innocent people who do fit the description, and $C = 1$ guilty person who fits the description. (The fourth possibility – a guilty person who doesn't fit the description – doesn't exist.) The two conditional probabilities that are relevant in the above discussion are then

$$P(D|I) = \frac{B}{\text{innocent}} = \frac{B}{B+A} = \frac{99}{999{,}999},$$

$$P(I|D) = \frac{B}{\text{fit description}} = \frac{B}{B+C} = \frac{99}{100}. \quad (3.244)$$

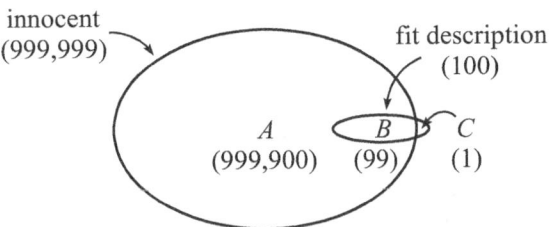

Figure 3.51

Both of these probabilities have B in numerator, because B represents the people who are innocent *and* fit the description. But the A in the first denominator is much larger than the C in second denominator. Or said in another way, B is a very small fraction of the innocent people (the large oval in Fig. 3.51), whereas it is a very large fraction of the people who fit the description (the small oval in Fig. 3.51).

The prosecutor's faulty reasoning has been used countless times in actual court cases, with tragic consequences. Innocent people have been convicted, and guilty people have walked free (the argument can work in that direction too). These consequences can't be blamed on the jury, of course. It is inevitable that many jurors will fail to spot the error in the reasoning. It would be silly to think that the entire population should be familiar with this issue in probability. Nor can the blame be put on the attorney making the argument. This person is either (1) overzealous and/or incompetent, or (2) entirely within his/her right to knowingly make an invalid argument (as distasteful as this may seem). In the end, the blame falls on either (1) the opposing attorney for failing to rebut the known logical fallacy, or (2) a legal system that in some cases doesn't allow a final rebuttal.

REMARK: In our solution above, we effectively used Bayes' theorem (a bread-and-butter tool in probability) without saying so. In one of its forms, Bayes' theorem states that for general events A and B,

$$P(A|B)P(B) = P(B|A)P(A). \qquad (3.245)$$

This equality follows from the fact that both sides are equal to $P(A$ and $B)$ (which is often denoted by $P(A \cap B)$). This in turn follows from the fact that the probability that both A and B occur equals the probability $P(B)$ that B occurs, multiplied by the fraction of those cases in which A also occurs (which is $P(A|B)$). Likewise with A and B switched. If we let A be I (innocent) and B be D (description) in Eq. (3.245), we obtain

$$P(I|D) = \frac{P(D|I) \cdot P(I)}{P(D)} = \frac{\frac{99}{999,999} \cdot \frac{999,999}{10^6}}{\frac{1}{10^4}} = \frac{99}{100}, \qquad (3.246)$$

as we found above. Alternatively, the numbers are a bit nicer if we work with G (for guilty) instead of I:

$$P(G|D) = \frac{P(D|G) \cdot P(G)}{P(D)} = \frac{1 \cdot \frac{1}{10^6}}{\frac{1}{10^4}} = \frac{1}{100}, \quad (3.247)$$

which correctly equals $1 - P(I|D)$. If we modify the problem and state that only one person in a million fits the description, then $P(D)$ in Eq. (3.247) becomes $1/10^6$, so $P(G|D) = 1$. This makes sense; only one person fits the description, so that person must be guilty. If we further modify the problem (keeping the $P(D) = 1/10^6$ change) and state that we're now dealing with New York (whose population we will take to be 10 million), then $P(G)$ in Eq. (3.247) becomes $1/10^7$, so $P(G|D) = 1/10$. Ten people now fit the description, one of whom is guilty.

Note that Bayes' theorem in Eq. (3.245) immediately tells us that the two conditional probabilities $P(A|B)$ and $P(B|A)$ are in general not equal. Their ratio is $P(A)/P(B)$, which in general isn't equal to 1.

Although our original Venn-diagram solution and our additional Bayes'-theorem solution are really the same thing in the end, my opinion is that drawing some Venn diagrams provides a more intuitive understanding than just plugging things into Bayes' theorem. ♣

47. **The game-show problem**

We'll present three solutions, one right and two wrong. You should decide which one you think is correct before reading beyond the third solution. Cover up the page after the third solution, so that you don't inadvertently see which one is correct.

- REASONING 1: Once the host reveals a goat, the prize must be behind one of the two remaining doors. Since the prize was randomly located to begin with, there must be equal chances that the prize is behind each of the two remaining doors. The probabilities are therefore both 1/2, so it doesn't matter if you switch.

    If you want, you can imagine a friend (who is aware of the whole procedure of the host announcing that he will open a door and reveal a goat) entering the room *after* the host opens the door. This person sees two identical unopened doors (he doesn't know which one you initially picked) and a goat. So for him there must be a 1/2 chance that the prize is behind each unopened door. The probabilities for you and your friend can't be any different, so you also say that each unopened door has a 1/2 chance of containing the prize. It therefore doesn't matter if you switch.

- REASONING 2: There is initially a 1/3 chance that the prize is behind any of the three doors. So if you don't switch, your probability of winning is 1/3. No actions taken by the host can change the fact that if you play a large number n of these games, then (roughly) $n/3$ of them will have the prize behind the door you initially pick.

Likewise, if you switch to the other unopened door, there is a 1/3 chance that the prize is behind that door. (There is obviously a goat behind at least one of the other two doors, so the fact that the host reveals a goat doesn't tell you anything new.) Therefore, since the probability is 1/3 whether or not you switch, it doesn't matter if you switch.

- REASONING 3: As in the first paragraph of Reasoning 2, if you don't switch, your probability of winning is 1/3.

  However, if you switch, your probability of winning is greater than 1/3. It increases to 2/3. This can be seen as follows. Without loss of generality, assume that you pick the first door. (You can repeat the following reasoning for the other doors if you wish. It gives the same result.) There are three equally likely possibilities for what is behind the three doors: PGG, GPG, and GGP, where P denotes the prize and G denotes a goat. If you don't switch, then in only the first of these three cases do you win, so your probability of winning is 1/3 (consistent with the first paragraph of Reasoning 2). But if you do switch from the first door to the second or third, then in the first case PGG you lose, but in the other two cases you win, because the door not opened by the host has the prize. (The host has no choice but to reveal the G and leave the P unopened.) Therefore, since two out of the three equally likely cases yield success if you switch, your probability of winning if you switch is 2/3. So you do in fact want to switch.

Which of these three solutions is correct? Don't read any further until you've firmly decided which one you think is right.

The third solution is correct. The error in the first solution is the statement, "there must be equal chances that the prize is behind each of the two remaining doors." This is simply not true. The act of revealing a goat breaks the symmetry between the two remaining doors, as explained in the third solution. One door is the one you initially picked, while the other door is one of the two that you didn't pick. The fact that there are two possibilities doesn't mean that their probabilities have to be equal, of course!

The error in the supporting reasoning with your friend (who enters the room after the host opens the door) is the following. While it *is* true that both probabilities are 1/2 for your friend, they aren't both 1/2 for *you*. The statement, "the probabilities for you and your friend can't be any different," is false. You have information that your friend doesn't have; you know which of the two unopened doors is the one you initially picked and which is the door that the host chose to leave unopened. (And as seen in the third solution, this information yields probabilities of 1/3 and 2/3.) Your friend doesn't have this critical information. Both doors look the same to him. Probabilities can certainly be different for different people. If I flip a coin and peek and see a Heads, but I don't show you, then the probability of a Heads is 1/2 for you, but 1 for me.

The error in the second solution is that the act of revealing a goat *does* give you new information, as we just noted. This information tells you that the prize isn't behind that door, and it also distinguishes between the two remaining unopened doors. One is the door you initially picked, while the other is one of the two doors that you didn't initially pick. As seen in the third solution, this information has the effect of increasing the probability that the goat is behind the other door. Note that another reason why the second solution can't be correct is that the two probabilities of 1/3 don't add up to 1.

To sum up, it should be no surprise that the probabilities are different for the switching and non-switching strategies *after* the host opens a door (the probabilities are obviously the same, equal to 1/3, whether or not a switch is made *before* the host opens a door), because the host gave you some of the information he had about the locations of things.

REMARKS:

1. If you still doubt the validity of the third solution, imagine a situation with 1000 doors containing one prize and 999 goats. After you pick a door, the host opens 998 other doors and reveals 998 goats (and he said beforehand that he was going to do this). In this setup, if you don't switch, your chances of winning are 1/1000. But if you do switch, your chances of winning are 999/1000, which can be seen by listing out (or imagining listing out) the 1000 cases, as we did with the three PGG, GPG, and GGP cases in the third solution. It is clear that the switch should be made, because the *only* case where you lose after you switch is the case where you had initially picked the prize, and this happens only 1/1000 of the time.

   In short, a huge amount of information is gained by the revealing of 998 goats. There is initially (and always) a 999/1000 chance that the prize is somewhere behind the other 999 doors, and the host is kindly giving you the information of exactly which door it is (in the highly likely event that it is in fact one of the other 999).

2. The clause in the statement of the problem, "The host announces that after you select a door (without opening it), he will open one of the other two doors and purposefully reveal a goat," is crucial. If it is omitted, and it is simply stated that "The host then opens one of the other two doors and reveals a goat," then it is impossible to state a preferred strategy. If the host doesn't announce his actions beforehand, then for all you know, he *always* reveals a goat (in which case you should switch, as we saw above). Or he *randomly* opens a door and just happened to pick a goat (in which case... well, you can think about that in Problem 48!). Or he opens a door and reveals a goat if and only if your initial door has the prize (in which case you definitely should not switch). Or he could have one procedure on Tuesdays and another on Fridays, each of which depends on the color of the socks he's wearing. And so on.

3. This problem is famous for the intense arguments it lends itself to. There is nothing terrible about getting the wrong answer, nor is there anything terrible about not believing the correct answer for a while. But concerning arguments that drag on and on, it doesn't make any sense to argue about this problem for more than, say, 20 minutes, because at that point everyone

should stop and just *play the game*! You can play a number of times with the switching strategy, and then a number of times with the non-switching strategy. Three coins with a dot on the bottom of one of them are all you need.[21] Not only will the actual game yield the correct answer (if you play enough times so that things average out), but the patterns that form will undoubtedly convince you of the correct reasoning (or reinforce it, if you're already comfortable with it). Arguing endlessly about an experiment, when you can actually *do* the experiment, is as silly as arguing endlessly about what's behind a door, when you can simply open the door.

4. For completeness, there is one subtlety we should mention here. In the second solution, we stated, "No actions taken by the host can change the fact that if you play a large number n of these games, then (roughly) $n/3$ of them will have the prize behind the door you initially pick." This part of the reasoning was correct; it was the "switching" part of the second solution that was incorrect. After doing Problem 48 (where the host randomly opens a door), you might disagree with the above statement, because it will turn out in that problem that the actions taken by the host *do* affect this $n/3$ result. However, the above statement is still correct for "*these*" games" (the ones governed by the original statement of the present problem). See the second remark in the solution to Problem 48 for further discussion. ♣

48. **A random game-show host**

    We'll solve this problem by listing out the various possibilities. Without loss of generality, assume that you pick the first door. (You can repeat the following reasoning for the other doors if you wish. It gives the same result.) There are three equally likely possibilities for what is behind the three doors: PGG, GPG, and GGP, where P denotes the prize and G denotes a goat. For each of these three possibilities, since you picked the first door, the host opens either the second or third door (with equal probabilities). So there are six equally likely results of his actions. These are shown in Table 3.6, with the bold letters signifying the object revealed.

	PGG	GPG	GGP
open 2nd door	PGG	G̶P̶G̶	GG**P**
open 3rd door	PGG	GP**G**	G̶G̶P̶

    Table 3.6: There are six equally likely scenarios with a randomly opened door, assuming that you pick the first door.

    We now note that the two results where the prize is revealed (the crossed-out GPG and GGP results) are not relevant to this problem, because we are told that the host happens to reveal a goat. Only the four other results are relevant:

---

[21] You actually don't need three objects. It's hard to find three exactly identical coins anyway. The "host" can simply roll a die, without showing the "contestant" the result. Rolling a 1 or 2 can mean that the prize is located behind the first door, a 3 or 4 the second, and a 5 or 6 the third. The game then basically involves calling out door numbers.

| PGG | PGG | GPG | GGP |

They are all still equally likely, so their probabilities must each be 1/4. We see that if you *don't* switch from the first door, you win on the first two of these results and lose on the second two. And if you *do* switch, you lose on the first two and win on the second two. So either way, your probability of winning is 1/2. It therefore doesn't matter if you switch.

REMARKS:

1. In the original setup in Problem 47, the probability of winning was 2/3 if you switched. How can it possibly decrease to 1/2 in the present random version, when in both versions the exact same thing happened, namely the host revealed a goat?

    The difference is due to the two cases where the host reveals the prize in the random version (the **GPG** and **GGP** cases). You don't benefit from these cases in the random version, because we are told in the statement of the problem that they don't exist. But in the original version, they represent guaranteed success if you switch, because the host is forced to open the other door, which is a goat.

    But still you may say, "If there are two setups, and if I pick, say, the first door in each, and if the host reveals a goat in each (by prediction in one case, and by random pick in the other), then *exactly the same thing happens in both setups.* How can the resulting probabilities (for winning on a switch) be different?" The answer is that although the two outcomes are the same, probabilities have nothing to do with *two* setups. Probabilities are defined only for a *large number* of setups. And if you play a large number of these pairs of games (prediction in one, random pick in the other), then in 1/3 of the pairs the host will reveal different things (a goat in the prediction version and the prize in the random version). These cases yield success in the original prediction version, but they are irrelevant in the random version. They are effectively thrown away there.

2. We will now address the issue mentioned in the fourth remark in the solution to Problem 47. We correctly stated that in the original version of the problem, "No actions taken by the host can change the fact that if you play a large number n of these games, then (roughly) $n/3$ of them will have the prize behind the door you initially pick." However, in the present random version of the problem, something *does* affect the probability that the prize is behind the door you initially pick. It is now 1/2 instead of 1/3. So can something affect this probability or not?

    Well, yes and no. If *all* of the n games are considered (as in the original version), then $n/3$ of them have the prize behind the initial door, and that's that. However, the random version of the problem involves throwing away 1/3 of the games (the ones where the host reveals the prize), because it is assumed in the statement of the problem that the host happens to reveal a goat. So for the *remaining games* (which are 2/3 of the initial total, hence $2n/3$), 1/2 of them ($n/3$ as always) have the prize behind your initial door.

If you play a large number n of games of each version (including the $n/3$ games that are thrown away in the random version), then the actual *number* of games that have the prize behind your initial door is the same, namely $n/3$. It's just that in the original version this number can be thought of as $1/3$ of n, whereas in the random version it can be thought of as $1/2$ of $2n/3$. So in the end, the thing that influences the probability (that the initial door you pick has the prize) and changes it from $1/3$ to $1/2$ isn't the opening of a door, but rather the throwing away of $1/3$ of the games. Since no games are thrown away in the original version, the above statement in quotes is correct (with the key phrase being "*these* games").

3. As with the original version of the problem, if you find yourself arguing about the answer for an excessive amount of time, you should just *play the game* (at least a few dozen times, to get good enough statistics). The randomness can be determined by a coin toss. As mentioned above, you will end up throwing away $1/3$ of the games (the ones where the host reveals the prize). ♣

49. The birthday problem

(a) There are many different ways for there to be *at least* one common birthday (one pair, two pairs, one triple, etc.), and it is completely intractable to add up all of these individual probabilities. It is *much* easier (and even with the italics, this is a vast understatement) to calculate the probability that there *isn't* a common birthday, and then subtract this from 1 to obtain the probability that there *is* at least one common birthday.

The calculation of the probability that there *isn't* a common birthday proceeds as follows. Let there be n people in the room. We can imagine taking them one at a time and randomly plopping their names down on a calendar, with the (present) goal being that there are no common birthdays. The first name can go anywhere. But when we plop down the second name, there are only 364 "good" days left, because we don't want the day to coincide with the first name's day. The probability of success for the second name is therefore $364/365$. Then, when we plop down the third name, there are only 363 "good" days left (assuming that the first two people have different birthdays), because we don't want the day to coincide with either of the other two days. The probability of success for the third name is therefore $363/365$. Similarly, the probability of success for the fourth name is $362/365$. And so on.

If there are n people in the room, the probability that all n birthdays are distinct (that is, there *isn't* a common birthday among any of the people; hence the superscript "no" below) therefore equals

$$P_n^{\text{no}} = 1 \cdot \frac{364}{365} \cdot \frac{363}{365} \cdot \frac{362}{365} \cdot \frac{361}{365} \cdots \cdot \frac{365-(n-1)}{365}. \tag{3.248}$$

If you want, you can write the initial 1 here as $365/365$, to make things look nicer. Note that the last term involves $(n-1)$ and not n, because $(n-1)$ is the number of names that have already been plopped down. As a double check that this $(n-1)$ is correct, it works for small numbers like $n = 2$ and 3. You should always perform a simple check like this whenever you write down *any* expression involving a parameter such as n.

We now just have to multiply out the product in Eq. (3.248) to the point where it becomes smaller than $1/2$, so that the probability that there *is* a common birthday is larger than $1/2$. With a calculator, this is tedious, but not horribly painful. We find that $P_{22}^{no} = 0.524$ and $P_{23}^{no} = 0.493$. If P_n^{yes} is the probability that there *is* a common birthday among n people, then $P_n^{yes} = 1 - P_n^{no}$, so $P_{22}^{yes} = 0.476$ and $P_{23}^{yes} = 0.507$. Since our original goal was to have $P_n^{yes} > 1/2$ (or equivalently $P_n^{no} < 1/2$), we see that there must be at least 23 people in a room in order for there to be a greater than 50% chance that at least two of them have the same birthday. The probability in the $n = 23$ case is 50.7%.

REMARK: The $n = 23$ answer to the problem is much smaller than most people would expect, so it provides a nice betting opportunity. For $n = 30$, the probability of a common birthday increases to 70.6%, and most people would still find it hard to believe that among 30 people, there are probably two who have the same birthday. Table 3.7 lists various values of n and the probabilities, $P_n^{yes} = 1 - P_n^{no}$, that at least two people have a common birthday. Even for $n = 50$, most people would probably be happy to bet, at even odds, that no two people have the same birthday. But you'll win the bet 97% of the time.

n	10	20	23	30	50	60	70	100
P_n^{yes}	11.7%	41.1%	50.7%	70.6%	97.0%	99.4%	99.92%	99.99997%

Table 3.7: Probability of a common birthday among n people.

Fig. 3.52 shows a plot of P_n^{yes}, for n from 1 to 70. (The short horizontal line associated with a given n has its left end at n and right end at $n + 1$.) The value of P_{23}^{yes} is just above the dashed horizontal line at height 0.5.

One reason why many people can't believe the $n = 23$ result is that they're asking themselves a different question, namely, "How many people (in addition to me) need to be present in order for there to be at least a $1/2$ chance that someone else has *my* birthday?" The answer to this question is indeed much larger than 23. The probability that *no one* out of n people has a birthday on a *given day* is simply $(364/365)^n$, because each person has a 364/365 chance of not having that particular birthday. For $n = 252$, this is just over $1/2$. And for $n = 253$, it is just under $1/2$; it equals 0.4995. Therefore, you need to come across 253 other people in order for the probability to be greater than $1/2$ that at least one of them *does* have *your* birthday (or any other particular birthday). ♣

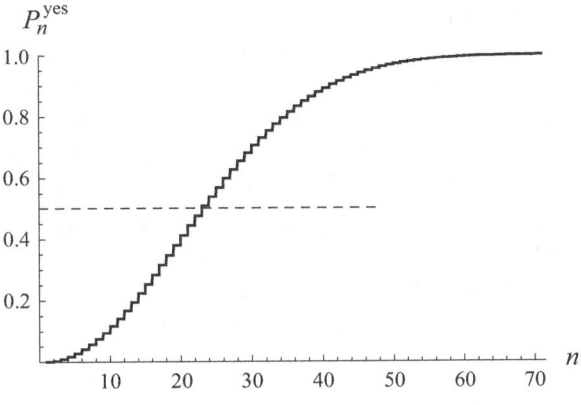

Figure 3.52

(b) FIRST SOLUTION: Given n people and N days in a year, the reasoning in part (a) tells us that the probability that no two people have the same birthday is

$$P_n^{no} = \left(1 - \frac{1}{N}\right)\left(1 - \frac{2}{N}\right) \cdots \left(1 - \frac{n-1}{N}\right). \quad (3.249)$$

It is often easier to work with the log of a product, so let's take the natural log of this equation and use the Taylor approximation,

$$\ln(1 - x) \approx -x - \frac{x^2}{2}. \quad (3.250)$$

(See the appendix for a review of Taylor series.) It turns out that we'll need only the first-order term here. But we'll include the second-order term, to show that it can in fact be ignored. Since the log of a product is the sum of the logs, the requirement $P_n^{no} < 1/2$ becomes

$$-\left(\frac{1}{N} + \frac{2}{N} + \cdots \frac{n-1}{N}\right) - \frac{1}{2}\left(\frac{1}{N^2} + \frac{4}{N^2} + \cdots \frac{(n-1)^2}{N^2}\right) < -\ln 2. \quad (3.251)$$

Using the sums,

$$\sum_1^m k = \frac{m(m+1)}{2} \quad \text{and} \quad \sum_1^m k^2 = \frac{m(m+1)(2m+1)}{6}, \quad (3.252)$$

we can rewrite Eq. (3.251) as

$$\frac{n(n-1)}{2N} + \frac{n(n-1)(2n-1)}{12N^2} > \ln 2. \quad (3.253)$$

For large N, the first term is of order 1 (which we need, in order for the inequality to hold) when $n \approx \sqrt{N}$, in which case the second term is negligible (being of order $N^{3/2}/N^2$). Therefore, keeping only the first term (which is essentially equal to $n^2/2N$, for large n), we find that P_n^{no} is equal to $1/2$ when

$$n \approx \sqrt{2\ln 2}\sqrt{N}. \quad (3.254)$$

Let's look at a few cases:
- For $N = 365$, Eq. (3.254) gives $n = 22.5$. Since we must have an integral number of people, this agrees with the exact result, $n = 23$.
- For $N = 24 \cdot 365 = 8760$ (that is, for births in the same hour), we find $n = 110.2$. This agrees with the exact result, $n = 111$, obtained by multiplying out Eq. (3.249) (not by hand on a calculator!).
- For $N = 60 \cdot 24 \cdot 365 = 525{,}600$ (that is, for births in the same minute), we find $n = 853.6$. This agrees with the exact result, $n = 854$, obtained by multiplying out Eq. (3.249). This is a very small number compared with the more than half a million minutes in a year.

REMARKS:

1. If we want to ask how many people need to be in a room in order for the probability to be at least p that two of them have the same birthday, then the above derivation is quickly modified to yield

$$n \approx \sqrt{2\ln\left(\frac{1}{1-p}\right)}\sqrt{N}. \qquad (3.255)$$

2. Recall the alternative question introduced in the remark in part (a): "How many people (in addition to me) need to be present in order for there to be at least a $1/2$ chance that someone else has *my* birthday?" What form does the answer take in the large-N limit? The probability that *no one* out of n people has a birthday on a given day is

$$\left(1 - \frac{1}{N}\right)^n \approx e^{-n/N}, \qquad (3.256)$$

where we have used Eq. (1.5) from Problem 53. This equals $1/2$ when $-n/N = \ln(1/2) \implies n = N\ln 2$. It is smaller than $1/2$ if $n > N\ln 2$. Therefore, if $n > N\ln 2$, you can expect that at least one of the n people *does* have your birthday. For $N = 365$, we find that $N\ln 2$ is slightly less than 253, so this agrees with the result obtained in the remark in part (a). Note that this $N\ln 2$ result is linear in N, whereas the result for the original problem in Eq. (3.254) behaves like $\sqrt{N}$. The reason for this square-root behavior can be seen in the following solution. ♣

SECOND SOLUTION: Given n people, there are $\binom{n}{2} = n(n-1)/2$ pairs. For large n, this is approximately equal to $n^2/2$. (The value of n we are concerned with will indeed turn out to be large, given our assumption of large N.) The probability that a given pair of people have the same birthday is $1/N$, so the probability that they do not have the same birthday is $1 - 1/N$. (This isn't quite correct for all of the pairs, because two pairs are not independent if, for example, they share a common person. But it is accurate enough for our purposes in the large-N limit. See the remark below.) Therefore, the probability that no pair has a common birthday is

$$P_n^{\text{no}} \approx \left(1 - \frac{1}{N}\right)^{n^2/2} \approx e^{-n^2/2N}, \qquad (3.257)$$

where we have used Eq. (1.5) from Problem 53. The righthand side equals 1/2 when

$$-n^2/2N = \ln(1/2) \implies n \approx \sqrt{2\ln 2}\,\sqrt{N}, \qquad (3.258)$$

in agreement with Eq. (3.254).

REMARK: (This is a long remark, so take a deep breath before diving in.) We assumed above that all of the pairs are independent, as far as writing down the $1 - 1/N$ probability goes. Let us now show that this is approximately true. We will show that for large N and n, the assumptions on the coincidence of birthdays in some pairs do not significantly affect the probability of coincidence in other pairs. More precisely, we will show that the relation $P_n^{\text{no}} \approx e^{-n^2/2N}$ in Eq. (3.257) is true if $n \ll N^{2/3}$, which covers the $n \propto N^{1/2}$ result in Eq. (3.258), assuming that N is very large.

Assumptions on the coincidence of birthdays in some pairs may slightly affect the probability of coincidence in other pairs, because the given assumptions may restrict the possible birthdays of the people in these other pairs. For example, if it is given that A and B do not have the same birthday, and also that B and C do not, then the probability that A and C *do* have the same birthday is 1/364 (which is larger than the naive 1/365), because A and C are both restricted from having a birthday on B's birthday, whatever it may be.

As another example, assume that A and B do not have the same birthday, and also that B and C do not, and also that C and D do not. What is the probability that A and D have the same birthday? For concreteness, assume that B's birthday is Jan 1, and C's birthday is Jan 2. (The exact days don't matter, as long as they are different.) Then A is restricted from having a birthday on Jan 1, and D is restricted from having a birthday on Jan 2. If A's birthday is Jan 2 (which occurs with probability 1/364), then there is a zero probability that D has the same birthday. For all the other 363 days (Jan 3 and onward), A and D each have a 1/364 chance of having that birthday, so the probability that A and D have the same birthday is $363 \cdot (1/364)^2$. This (only slightly) smaller than the naive 1/365.

The preceding two paragraphs show that when restrictions due to other pairs are taken into account, the probability that a given pair has a common birthday might be larger or smaller than the naive 1/365.

Given n people and a large number N of days in a year, our strategy will be to produce upper and lower bounds on the probability that two people have the same birthday, taking restrictions into account. We will then show that these bounds are close enough to $1/N$ so that the difference from $1/N$ can be ignored. Our bounds will be very generous; we will make no attempt at determining the actual attainable bounds. Our only goal will be to produce bounds that are sufficient for our purposes. The reasoning is as follows.

The probability that a given pair of people have the same birthday is

$$P = \sum_{i=1}^{N} p_1(d_i)\,p_2(d_i), \qquad (3.259)$$

where $p_1(d_i)$ is the probability that the first person in the pair has a birthday on day d_i, and likewise for $p_2(d_i)$ and the second person. The sum runs over all of the N days. If there were no restrictions due to other pairs, then every $p(d)$ would simply be $1/N$. And since there are N terms in the sum, we would obtain $P = N \cdot (1/N)^2 = 1/N$. This is correctly the naive probability that two people have the same birthday, ignoring any restrictions.

But there are restrictions. Let's produce an upper bound on P for any given pair, for any set of restrictions. Consider one pair, and assume that the other $n - 2$ birthdays have been chosen. An upper bound on the value that $p_1(d_i)$ or $p_2(d_i)$ can take is $1/(N - (n - 2))$, because a given person can be restricted from at most (depending on how many pairs are already specified) the $n - 2$ birthdays of the other $n - 2$ people. And since there is nothing to distinguish the remaining $N - (n-2)$ days, they all have equal probabilities of $1/(N - (n-2))$. For simplicity, we'll replace this with the larger number $1/(N - n)$, which is therefore also an (unreachable) upper bound.

Now, it certainly can't be the case that all $p(d)$ values take on (or are close to) the upper bound of $1/(N - n)$, because many of the $p(d)$'s are zero, due to the restrictions. But we're just trying to produce an upper bound on P here, so we'll simply set all the $p(d)$'s in Eq. (3.259) equal to $1/(N - n)$. Since there are N terms in the sum, an (unreachable) upper bound on P is therefore

$$P_{\text{upper}} = N \cdot \frac{1}{(N - n)^2}. \tag{3.260}$$

The n's we will be concerned with are much smaller than N, so to leading order in N, you can show that (using $1/(1 - \epsilon) \approx 1 + \epsilon$)

$$P_{\text{upper}} \approx \frac{N + 2n}{N^2}. \tag{3.261}$$

It actually isn't necessary to perform this approximation at this point, but it makes the math below a bit cleaner. If you're worried about ruining the upper bound, you can just change the $2n$ to $3n$, which will safely maintain an upper bound, assuming that N is large (which in turn will end up implying that it is much larger than n).

Let's now produce a lower bound on P. Some of the $p(d)$'s may be zero, while many are not. Given a particular set of specified birthdays, the smallest that any nonzero $p(d)$ can be is $1/N$, which occurs in the case of no restrictions. How many $p(d)$'s can be zero? Well, at most (depending on how many pairs are already specified) $n - 2$ of the $p_1(d_i)$'s can be zero, because there are at most $n - 2$ other birthdays that can be specified. Likewise, at most $n - 2$ of the $p_2(d_i)$'s can be zero. Now, there may very well be some overlap in the d_i days in these two sets (complete overlap if all the other pairs are specified). But since we're just trying to get a lower bound on P, let's assume that there is no overlap, which means that up to $2 \cdot (n - 2)$ terms in the sum in Eq. (3.259) can be zero. Let's increase this to $2n$ for simplicity, which makes P even smaller. Then at most $2n$ of the terms in the sum are zero, and at least $N - 2n$ of the terms are at least

$(1/N) \cdot (1/N)$. An (unreachable) lower bound on P is therefore

$$P_{\text{lower}} = 2n \cdot 0 + (N - 2n) \cdot \frac{1}{N^2} = \frac{N - 2n}{N^2}. \tag{3.262}$$

The probability that a given pair does *not* have a common birthday is $1 - P$. So a lower bound on this probability is $1 - P_{\text{upper}}$, and an upper bound is $1 - P_{\text{lower}}$. Therefore, instead of Eq. (3.257), we now have

$$\left(1 - P_{\text{upper}}\right)^{n^2/2} < P_n^{\text{no}} < \left(1 - P_{\text{lower}}\right)^{n^2/2}$$

$$\implies \left(1 - \frac{N + 2n}{N^2}\right)^{n^2/2} < P_n^{\text{no}} < \left(1 - \frac{N - 2n}{N^2}\right)^{n^2/2}$$

$$\implies e^{-n^2(N+2n)/2N^2} < P_n^{\text{no}} < e^{-n^2(N-2n)/2N^2}, \tag{3.263}$$

where we have used Eq. (1.5) from Problem 53. The ratio of these upper and lower bounds on P_n^{no} is

$$\exp\left(-\frac{n^2(N - 2n)}{2N^2} + \frac{n^2(N + 2n)}{2N^2}\right) = \exp\left(\frac{4n^3}{2N^2}\right). \tag{3.264}$$

This ratio is essentially equal to 1 (in which case P_n^{no} is squeezed down to the value in Eq. (3.257)), provided that $n^3 \ll N^2 \implies n \ll N^{2/3}$. Therefore, $P_n^{\text{no}} \approx e^{-n^2/2N}$ if $n \ll N^{2/3}$. And since the $n \propto N^{1/2}$ result in Eq. (3.258) is in this realm (assuming N is large), it is therefore valid. ♣

EXTENSION: We can also ask the following question: How many people need to be in a room in order for there to be a greater than $1/2$ probability that at least b of them have the same birthday? (So $b = 2$ corresponds to the case of pairs we solved above.) Assume that there is a very large number N of days in a year, and ignore effects that are of subleading order in N. (Think about this before reading further.)

We can solve this problem in the manner of the second solution above. Given b people, there are $\binom{n}{b} = (n!/(n - b)!)(1/b!)$ groups of b people. For large n, this is approximately equal to $n^b/b!$, assuming $b \ll n$. (As with the case of pairs above, n will indeed be large here.) The probability that a given group of b people all have the same birthday is $1/N^{b-1}$, so the probability that they do not all have the same birthday is $1 - 1/N^{b-1}$. (Again, this isn't quite correct, but it's close enough. See the first remark below.) Therefore, the probability, P_n^{no}, that no group of b people all have the same birthday is

$$P_n^{\text{no}} \approx \left(1 - \frac{1}{N^{b-1}}\right)^{n^b/b!} \approx e^{-n^b/b!N^{b-1}}, \tag{3.265}$$

where we have again used Eq. (1.5) from Problem 53. The righthand side equals $1/2$ when

$$n \approx (b! \ln 2)^{1/b} N^{1-1/b}. \tag{3.266}$$

REMARKS:

1. As with the case of pairs, we can show (for large N and n) that assumptions on the coincidence of birthdays in some b-tuples do not significantly affect the probability of coincidence in other b-tuples. We'll just sketch the reasoning here; you can fill in the gaps. Consider $b = 3$ for concreteness; larger b values are treated similarly. In place of Eq. (3.259), the probability that a given triplet of people all have the same birthday is

$$P = \sum_{i=1}^{N} p_1(d_i) p_2(d_i) p_3(d_i). \qquad (3.267)$$

From the same type of reasoning as in the case of pairs, you can convince yourself that a (very generous) upper bound on P is $P_{\text{upper}} = N/(N-n)^3 \approx (N+3n)/N^3$. And a (very generous) lower bound on P is $P_{\text{lower}} = (N-3n)/N^3$.[22] With $b = 3$ in (the modification of) Eq. (3.265), the bounds analogous to those in Eq. (3.263) take the form,

$$e^{-n^3(N+3n)/6N^3} < P_n^{\text{no}} < e^{-n^3(N-3n)/6N^3}. \qquad (3.268)$$

The ratio of these bounds is $e^{6n^4/6N^3}$. This ratio is essentially equal to 1, provided that $n^4 \ll N^3 \implies n \ll N^{3/4}$. Therefore, the $P_n^{\text{no}} \approx e^{-n^3/6N^2}$ result in Eq. (3.265) (with $b = 3$) is valid if $n \ll N^{3/4}$. And since the $n \propto N^{2/3}$ result in Eq. (3.266) (with $b = 3$) is in this realm (assuming N is large), it is therefore valid.

2. Eq. (3.266) holds in the large-N limit. If we wish to make another approximation, that of large b, we can say $(b! \ln 2)^{1/b} \approx b/e$. (This follows from Stirling's formula, $m! \approx m^m e^{-m} \sqrt{2\pi m}$.) Therefore, for large N, n, and b (with $b \ll n \ll N$), we have $P_n^{\text{no}} = 1/2$ when

$$n \approx (b/e) N^{1-1/b}. \qquad (3.269)$$

3. The right-hand side of equation Eq. (3.266) scales with N according to $N^{1-1/b}$. This means that if we look at the numbers of people needed to have a greater than $1/2$ chance that pairs, triplets, etc., have common birthdays, we see that these numbers scale like

$$N^{1/2}, N^{2/3}, N^{3/4}, \cdots. \qquad (3.270)$$

For large N, these results are multiplicatively far apart. Therefore, there are values of n for which we can say, for example, that we are virtually certain that there are pairs and triplets with common birthdays, but also that we are virtually certain that there are no quadruplets with a common

---

[22]Neglecting terms of order 1, the $3n$ (or more generally bn) terms in these two bounds are really $bn/(b-1)$. The b in the numerator comes from the fact that there are b of the $p(d)$ factors in the generalization of Eq. (3.267). And the $b-1$ in the denominator comes from the fact that a birthday is ruled out only if $b-1$ other people have that birthday; so at most (roughly) $n/(b-1)$ days can be ruled out. The safer bounds obtained by dropping the $b-1$ are still sufficient for our purposes.

birthday. For example, if $n = N^{17/24}$ (which satisfies $N^{2/3} < n < N^{3/4}$), Eq. (3.265) tells us that the probability that there is a common birthday triplet is $1 - e^{-(1/6)N^{1/8}} \approx 1$, whereas the probability that there is a common birthday quadruplet is $1 - e^{-(1/24)N^{-1/6}} \approx (1/24)N^{-1/6} \approx 0$, where we have used $e^{-x} \approx 1 - x$, for small x. ♣

50. **The boy/girl problem**

    (a) The key to all three formulations of the problem is to list out the various equally likely possibilities for the family's children, while taking into account only the "I have two children" information, and *not yet* the information about the boy. With B for boy and G for girl, the family in the present scenario in part (a) can be of four types (at least *before* the parent gives you information about the boy), each with probability 1/4:

    $$\boxed{\text{BB}} \quad \boxed{\text{BG}} \quad \boxed{\text{GB}} \quad \text{GG}$$

    Ignore the boxes for a moment. In each pair of letters, the first letter stands for the older child, and the second letter stands for the younger child. We could just as well order them by, say, height or shoe size, but the ordering by age will be convenient for part (b).

    Note that there are indeed four equally likely possibilities (BB, BG, GB, GG), as opposed to just three equally likely possibilities (BB, BG, GG), because the older child has a 50-50 chance of being a boy or a girl, as does the younger child. The BG and GB cases each get counted once (instead of being lumped together), just as the HT and TH cases each get counted once when flipping two coins, where the four equally likely possibilities are HH, HT, TH, TT.

    Under the assumption of general randomness stated in the problem, we are assuming that you are equally likely (at least *before* the parent gives you information about the boy) to bump into a parent of any one of the above four types of two-child families.

    Let us *now* invoke the information that at least one child is a boy. This information tells us that you can't be talking with a GG parent. The parent must be a BB, BG, or GB parent, all equally likely. (They are equally likely, because they are all equivalent with regard to the "at least one of them is a boy" statement.) These are the boxed families in the above list. Of these three cases, only the BB case has the other child being a boy. The desired probability that the other child is a boy is therefore 1/3.

    If you don't trust the reasoning in the preceding paragraph, just imagine performing many trials of the setup. This is always a good strategy when solving probability problems. Imagine that you encounter 1000 random parents of two children. You will encounter about 250 of each of the four types of parents. The 250 GG parents have nothing to do with the given setup, so we must discard them. Only the other 750 parents (BB, BG, GB) are able to provide the given information that at least one child is a boy. Of these 750 parents, 250 are of the BB type and thereby have a boy as the other child. The desired probability is therefore 250/750 = 1/3.

(b) As in part (a), *before* the information about the boy is taken into account, there are four equally likely possibilities for the children (again ignore the boxes for a moment):

$$\boxed{\text{BB}} \quad \boxed{\text{BG}} \quad \text{GB} \quad \text{GG}$$

But once the parent tells you that the older child is a boy, the GB and GG cases are ruled out; remember that the first letter in each pair corresponds to the older child. So you must be talking with a BB or BG parent, both equally likely. Of these two cases, only the BB case has the other child being a boy. The desired probability that the other child is a boy is therefore 1/2.

(c) This version of the problem is a little trickier, because there are now *eight* equally likely possibilities (before the information about the boy is taken into account), instead of just four. This is true because for each of the four types of families in the above lists, either of the children may be chosen to go on the walk (with equal probabilities, as we are assuming for everything). The eight equally likely possibilities are therefore shown in Table 3.8; again ignore the boxes for a moment. The bold letter indicates the child you encounter. (And the first letter still corresponds to the older child.)

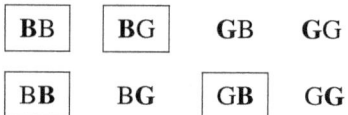

Table 3.8: The eight types of families, accounting for the child present.

Once the parent tells you that one of the children is the boy standing there, four of the eight possibilities are ruled out. Only the four boxed pairs in Table 3.8 (the ones with a bold **B**) satisfy the condition that the child standing there is a boy. Of these four (equally likely) possibilities, two of them have the other child being a boy. The desired probability that the other child is a boy is therefore 1/2.

REMARKS:

1. We used the given assumption of general randomness many times in the above solutions. One way to make things *non*random is to assume that the parent who is out for a walk is chosen randomly with equal 1/3 probabilities of being from BB families, or GG families, or one-boy-and-one-girl families. This is an artificial construction, because it means that a given BG or GB family (which together make up half of all two-child families) is less likely to be chosen than a given BB or GG family. This violates our assumption of general randomness. In this scenario, you can show that the answers to parts (a), (b), and (c) are 1/2, 2/3, and 2/3.

Another way to make things nonrandom is to assume that in part (c) a girl is always chosen to go on the walk if the family has at least one girl. The answer to part (c) is then 1, because the only way a boy will be standing there is if both children are boys. On the other hand, if we assume that a boy is always chosen to go on the walk if the family has at least one boy, then the answer to part (c) is 1/3. This is true because for BB, the other child is a boy; and for both BG and GB (for which the boy is always chosen to go on the walk), the other child is a girl. Basically, the middle four pairs in Table 3.8 will all have a bold **B**, so they will all be boxed. There are countless ways to make things nonrandom, so unless we make an assumption of general randomness, there is no way to solve the problem (without specifying exactly what the nonrandomness is).

2. Let's compare the scenarios in parts (a) and (b), to see exactly why the probabilities differ. In part (a), the parent's statement rules out the GG case. The BB, BG, and GB cases survive, with the BB families representing 1/3 of all of the possibilities. If the parent then changes the statement "at least one of them is a boy" to "the older one is a boy," we are now in the realm of part (b). The GB case is now also ruled out (in addition to the GG case). So only the BB and BG cases survive, with the BB families representing 1/2 of all of the possibilities. This is why the probability jumps from 1/3 to 1/2 in going from part (a) to part (b). An additional group of families (GB) is ruled out.

   Let's now compare the scenarios in parts (a) and (c), to see exactly why the probabilities differ. As in the preceding paragraph, the parent's statement in part (a) rules out the GG case. If the parent then makes the additional statement, "... and there he is over there next to that tree," we are now in the realm of part (c). Which additional families are ruled out? Well, in part (a), you could be talking with a parent in any of the families in Table 3.8 except the two GG entries. So there are six possibilities. But as soon as the parent adds the "and there he is" comment, the unboxed **GB** and **BG** entries are ruled out. So a larger fraction of the possibilities (now two out of four, instead of two out of six) have the other child being a boy.

3. Having gone through all of the above reasonings and the comparisons of the different cases, we should note that there is actually a much quicker way of obtaining the probabilities of 1/2 in parts (b) and (c). If the parent says that the older child is a boy, or that one of the children is the boy standing next to her, then the parent is making a statement *solely about a particular child* (the older one, or the present one). The parent is saying nothing about the other child (the younger one, or the absent one). We therefore know nothing about that child. So by our assumption of general randomness, the other child is equally likely to be a boy or a girl. This should be contrasted with part (a). In that scenario, when the parent says that at least one child is a boy, the parent is *not* making a claim about a *specific* child, but rather about the *collective set* of the two children together. We are therefore not able to uniquely define the "other child" and simply say that the answer is 1/2. The answer depends on both children together, and it turns out to be different from 1/2 (namely 1/3).

4. As a generalization to part (a), we can change the person's statement to: "I have N children. At least $N-1$ of them are boys." What is the probability that all N children are boys? Let's consider the $N = 3$ case. Before the information about the boys is taken into account, there are $2^3 = 8$ equally likely possibilities for the children: BBB, BBG, BGB, GBB, BGG, GBG, GGB, GGG (ordered by age, shoe size, or whatever). The additional information that at least $N-1 = 2$ of the children are boys then leaves us with BBB, BBG, BGB, and GBB (all equally likely). The desired probability that all three children are boys is therefore $1/4$. For a general N, analogous reasoning quickly gives the answer of $1/(N+1)$.

In short, with N children, there is only one way to have N boys, but N ways to have $N-1$ boys and one girl. The same reasoning applies when flipping coins. For example, if you flip ten coins, there is a $1/2^{10} = 1/1024$ chance of getting all ten Heads, and a $10/1024$ chance of getting nine Heads and one Tails, because there are ten possibilities for which coin is Tails. (The coins can be ordered according to, say, shininess.) So if I flip ten coins and tell you that I got at least nine Heads, there is only a $1/11$ chance that all ten coins are Heads.[23]

5. If you're not convinced of the answer of $1/3$ for part (a), or if you find yourself arguing about it with someone for more than 20 minutes, then you can simply play the same. You just need to (repeatedly) flip two coins. Heads is a boy, Tails is a girl. After each flip of the pair of coins, ask yourself: Is at least one child a boy? If so, check if the other child is also a boy. Do this many times. You will find that the other child is a boy in (roughly) $1/3$ of the trials.

6. There is a subtlety in this problem that we should address: How does the parent decide what information to give you? A reasonable rule could be that in part (a) the parent says, "At least one child is a boy," if she is able to; otherwise she says, "At least one child is a girl." This is consistent with all of our above reasoning. But consider what happens if we tweak the rule so that now the parent says, "At least one child is a girl," if she is able to; otherwise she says, "At least one child is a boy." In this case, the answer to part (a) is 1, because the only parents making the "boy" statement are the BB parents. This minor tweak completely changes the problem.

If you want to avoid this issue, you can rephrase part (a) as: You bump into a random person on the street and ask, "Do you have (exactly) two children? If so, is at least one of them a boy?" In the cases where the answers to both of these questions are "yes," what is the probability that the other child is also a boy? Alternatively, you can just remove the parent and pose the problem as: Consider all two-child families that have at least one boy. What is the probability that both children are boys? This phrasing isn't as catchy as the original, but it gets rid of the above issue.

---

[23] This is the probability from your point of view. From *my* point of view, the probability that all ten coins are Heads is simply either 0 or 1, because I can see all of the coins; everything is determined. This distinction applies to the original problem too, because the parent of course knows the nature of the children. The probabilities we have been talking about throughout this problem are from your point of view, based on the information you have (which is less than what the parent has).

7. In the various lists of types of families in the above solutions, only the boxed types were applicable. The unboxed ones didn't satisfy the conditions given in the statement of the problem, so we discarded them. This act of discarding the unboxed types is equivalent to using the conditional-probability statement, $P(A \text{ and } B) = P(A) \cdot P(B|A)$, where $P(B|A)$ stands for the probability that B occurs, given that A occurs. (This relation is true because we can identify the events where both A and B occur by looking at all the events where A occurs and then looking at the fraction of these where B also occurs.) We can rearrange the relation to give

$$P(B|A) = \frac{P(A \text{ and } B)}{P(A)}. \qquad (3.271)$$

(This is a minimalistic form of Bayes' theorem.) In part (a), if we let $A = \{\text{at least 1 boy}\}$ and $B = \{2 \text{ boys}\}$, then we obtain

$$P\big((2 \text{ boys})|(\text{at least 1 boy})\big) = \frac{P\big((\text{at least 1 boy}) \text{ and } (2 \text{ boys})\big)}{P(\text{at least 1 boy})}. \qquad (3.272)$$

The lefthand side of this equation is the probability we're trying to find. On the righthand side, we can rewrite $P((\text{at least 1 boy}) \text{ and } (2 \text{ boys}))$ as just $P(2 \text{ boys})$, because $\{2 \text{ boys}\}$ is a subset of $\{\text{at least 1 boy}\}$. So we have

$$P\big((2 \text{ boys})|(\text{at least 1 boy})\big) = \frac{P(2 \text{ boys})}{P(\text{at least 1 boy})} = \frac{1/4}{3/4} = \frac{1}{3}. \qquad (3.273)$$

The preceding equations might look a bit intimidating, which is why we took a more intuitive route in our original solution to part (a), where we wrote out the possibilities and then boxed the relevant (non-GG) ones, or where we imagined doing 1000 trials and then discarding the 250 GG families. Discarding these families accomplishes the same thing as having the $P(\text{at least 1 boy})$ term in the denominator in Eq. (3.273); namely, they both signify that we are concerned only with families that have at least one boy.

8. If you thought that some of the answers to this problem were counterintuitive, then, well, you haven't seen anything yet! Tackle Problem 51 and you'll see why. ♣

51. Boy/girl problem with general information

Let's be general right from the start and consider the case where the boy has a particular characteristic that occurs with probability p. (So $p = 1/4$ if the characteristic is a summer birthday.) As in all of the versions in Problem 50, we'll list out the various possibilities in a table, *before* the parent's additional information (beyond "I have two children") is taken into account. It is still the case that the BB, BG, GB, and GG types of two-child families are all equally likely, with a 1/4 probability for each. We are again ordering the children in a given pair by age; the first letter is associated with the older child. But we could just as well order them by, say, height or shoe size.

In the present version of the problem, there are now various different subtypes within each type of family, depending on whether or not the children have the given characteristic (which occurs with probability p). For example, if we look at the BB types, there are four possibilities for the occurrence(s) of the characteristic. With "y" standing for "yes, the child has the characteristic," and "n" standing for "no, the child doesn't have the characteristic," the four possibilities are B_yB_y, B_yB_n, B_nB_y, and B_nB_n. (In the second possibility here, for example, the older boy has the characteristic, and the younger boy doesn't.) Since y occurs with probability p, we know that n occurs with probability $1-p$. The probabilities associated with each of the four possibilities are therefore equal to the 1/4 probability that BB occurs, multiplied by, respectively, p^2, $p(1-p)$, $(1-p)p$, and $(1-p)^2$.

The same reasoning holds with the BG, GB, and GG types, so we obtain a total of $4 \cdot 4 = 16$ distinct possibilities. These are listed in Table 3.9 (ignore the boxes for a moment). The four subtypes in any given row all have the same occurrence(s) of the characteristic, so they all have the same probability; this probability is listed on the right. The subtypes in the middle two rows all have equal probabilities. As mentioned above, in the case where the given characteristic is "having a birthday in the summer," p equals 1/4. So the probabilities associated with the four rows in that case are equal to 1/4 multiplied by, respectively, 1/16, 3/16, 3/16, and 9/16.

	BB	BG	GB	GG	Probability
yy	B_yB_y	B_yG_y	G_yB_y	G_yG_y	$(1/4) \cdot p^2$
yn	B_yB_n	B_yG_n	G_yB_n	G_yG_n	$(1/4) \cdot p(1-p)$
ny	B_nB_y	B_nG_y	G_nB_y	G_nG_y	$(1/4) \cdot p(1-p)$
nn	B_nB_n	B_nG_n	G_nB_n	G_nG_n	$(1/4) \cdot (1-p)^2$

Table 3.9: The 16 types of families.

Before the parent gives you the additional information, all 16 of the subtypes in the table are possible. But after the statement is made that there is at least one boy with the given characteristic (that is, there is at least one B_y in the pair of children), only seven subtypes remain. These are indicted with boxes. The other nine subtypes are ruled out.

We now simply observe that the three boxes in the left-most column in the table have the other child being a boy, while the four other boxes in the second and third columns have the other child being a girl. The desired probability that the other child is a boy is therefore equal to the sum of the probabilities of the left three boxes, divided by the sum of the probabilities of all seven boxes. This gives

(ignoring the common factor of $1/4$ in all of the probabilities)

$$P_{BB} = \frac{p^2 + 2\cdot p(1-p)}{3\cdot p^2 + 4\cdot p(1-p)} = \frac{2p - p^2}{4p - p^2} = \frac{2-p}{4-p}. \qquad (3.274)$$

In the case where the given characteristic is "having a birthday in the summer," p equals $1/4$. Plugging this into Eq. (3.274) tells us that the probability that the other child is also a boy is $P_{BB} = 7/15 = 0.467$.

If the given characteristic is "having a birthday on August 11th," then $p = 1/365$, which yields $P_{BB} = 729/1459 = 0.4997 \approx 1/2$.

If the given characteristic is "being born during a particular minute on August 11th," then p is essentially equal to zero, so Eq. (3.274) tells us that P_{BB} is essentially equal to $1/2$. This makes sense, because if $p = 0$, the $p(1-p)$ probability for the middle two rows in Table 3.9 is much larger than the p^2 probability for the top row. Of course, *all* of these probabilities (in the first three rows) are very small in the small-p limit, but p^2 is much smaller than $p(1-p) \approx p$ when p is small. So we can ignore the top row. We are then left with four boxes, two of which are BB and two of which are BG/GB. The desired probability is therefore $1/2$.

Another somewhat special case is $p = 1/2$. (You can imagine that every child flips a coin, and we're concerned with the children who get Heads.) In this case we have $p = 1 - p$, so all of the probabilities in the righthand column in Table 3.9 are equal. All 16 entries in the table therefore have equal probabilities (namely $1/16$). Determining probabilities is then just a matter of counting boxes, so the answer to the problem is $3/7$, because three of the seven boxes are of the BB type.

REMARKS:

1. The above $P_{BB} \approx 1/2$ result in the $p \approx 0$ case leads to the following puzzle. Let's say that you bump into a random person on the street who says, "I have two children. At least one of them is a boy." At this stage, you know that the probability that the other child is also a boy is $1/3$, from part (a) of the original setup in Problem 50. But if the parent then adds, "... who was born during a particular minute on August 11th," then we just found that the probability that the other child is also a boy jumps to (essentially) $1/2$. Why exactly did this jump take place?

   In part (a) of Problem 50, there were three equally likely possibilities after the parent gave the additional information, namely BB, BG, and GB. Only $1/3$ of these cases (namely BB) had the other child being a boy. In the new scenario (with $p \approx 0$), there are four equally likely possibilities after the parent gives the additional information, namely B_yB_n, B_nB_y, B_yG_n, and G_nB_y. (As mentioned above, we're ignoring the top row in Table 3.9 since $p \approx 0$.) So in the new scenario, $1/2$ of these cases (the two BB cases) have the other child being a boy. The critical point here is that BB now counts *twice*, whereas it counted only once in the original scenario. This is due to the fact that a BB parent is (essentially) *twice as likely* (compared with a BG or GB parent) to be able to say that a boy was born during a particular minute on August 11th, because with two boys there are two chances to achieve this highly improbable characteristic. In contrast, a BB parent is *no more likely* (compared with a BG or GB parent) to be able to say simply that at least one child is a boy.

2. In the other extreme where the given characteristic is "being born on *any* day," we have $p = 1$. (This clearly isn't much of a characteristic, since it is satisfied by everyone.) So Eq. (3.274) gives $P_{BB} = 1/3$. In this $p = 1$ case, only the entries in the top row in Table 3.9 have nonzero probabilities. We are therefore in the realm of the first scenario in Problem 50, where we started off with the four types of families (BB, BG, GB, GG) and then ruled out the GG type, yielding a probability of $1/3$. It makes sense that the $1/3$ answer in the $p = 1$ case is the same as the $1/3$ answer in the first scenario in Problem 50, because the "being born on *any* day" statement provides no additional information. So the setup is equivalent to the first scenario in Problem 50, where the parent provided no additional information (beyond the fact that at least one child was a boy). ♣

52. **Stirling's formula**

(a) Let's first prove the result,

$$N! = \int_0^\infty x^N e^{-x} dx. \quad (3.275)$$

The proof by induction proceeds as follows. Integrating by parts gives

$$\int_0^\infty x^N e^{-x} dx = -x^N e^{-x} dx \Big|_0^\infty + N \int_0^\infty x^{N-1} e^{-x} dx. \quad (3.276)$$

The first term on the righthand side is zero, so if we define the integral in Eq. (3.275) as I_N, then Eq. (3.276) gives $I_N = NI_{N-1}$. Therefore, if $I_{N-1} = (N-1)!$, then $I_N = N!$. Since it is indeed true that $I_0 \equiv \int_0^\infty e^{-x} dx = -e^{-x}\Big|_0^\infty = 1 = 0!$, we see that $I_N = N!$ for all N.

We'll now follow the given hint and write $x^N e^{-x}$ as $e^{N \ln x - x} \equiv e^{f(x)}$, and then expand $f(x)$ in a Taylor series about its maximum. The maximum of $f(x)$ occurs at $x = N$ because $f'(x) = N/x - 1$, which equals zero when $x = N$.

To determine the Taylor series expanded around $x = N$, we must take derivatives of $f(x)$ and evaluate them at $x = N$; see Eq. (4.1) in the appendix, with $x_0 \equiv N$. We'll need to go only to second order here. We already know that the first derivative, $f'(x) = N/x - 1$, is zero at $x = N$. The second derivative is $f''(x) = -N/x^2$, which takes the value of $-1/N$

at $x = N$. The desired Taylor series is therefore

$$f(x) \approx f(N) + f'(N)(x-N) + \frac{f''(N)}{2!}(x-N)^2$$
$$= (N \ln N - N) + 0 - \frac{(x-N)^2}{2N}. \tag{3.277}$$

Plugging this expression for $f(x)$ into Eq. (3.275) gives

$$N! = \int_0^\infty e^{f(x)} \, dx \approx \int_0^\infty \exp\left(N \ln N - N - \frac{(x-N)^2}{2N}\right) dx$$
$$= N^N e^{-N} \int_0^\infty \exp\left(-\frac{(x-N)^2}{2N}\right) dx. \tag{3.278}$$

If N is very large, we can let the integral run from $-\infty$ to ∞, with negligible error. This is true because at $x = 0$, the value of the integrand is $e^{-N/2}$, which is essentially zero when N is large. The integrand is even smaller for negative values of x, which therefore contribute negligibly to the integral. Letting $y \equiv x - N$ in Eq. (3.278) then gives

$$N! \approx N^N e^{-N} \int_{-\infty}^\infty e^{-y^2/2N} \, dy$$
$$= N^N e^{-N} \sqrt{2\pi N}, \tag{3.279}$$

as desired. We have used the fact that $\int_{-\infty}^\infty e^{-x^2/b} \, dx = \sqrt{b\pi}$. This can be proved in the following way, where we make use of a change of variables from Cartesian to polar coordinates. Let $I \equiv \int_{-\infty}^\infty e^{-x^2} \, dx$. Then $I = \sqrt{\pi}$, because

$$I^2 = \int_{-\infty}^\infty e^{-x^2} \, dx \int_{-\infty}^\infty e^{-y^2} \, dy$$
$$= \int_{-\infty}^\infty \int_{-\infty}^\infty e^{-(x^2+y^2)} \, dx \, dy$$
$$= \int_0^{2\pi} \int_0^\infty e^{-r^2} r \, dr \, d\theta$$
$$= 2\pi \left(-\frac{e^{-r^2}}{2}\right)\Bigg|_0^\infty$$
$$= \pi. \tag{3.280}$$

(To go from the second to third line, we used the fact that the area element in polar coordinates is $r \, dr \, d\theta$. The limits correspond to the range of integration being the entire plane.) A change of variables with $x \equiv y/\sqrt{b}$ then turns $\int_{-\infty}^\infty e^{-x^2} \, dx = \sqrt{\pi}$ into $\int_{-\infty}^\infty e^{-y^2/b} \, dy = \sqrt{b\pi}$, as we wanted to show.

REMARKS:

1. Stirling's formula is a good approximation to $N!$, in the sense that the ratio of the approximate value to the true value approaches 1 for large N. (Even for just $N = 10$, the error is only about 1%.) An equivalent way of saying this is that *multiplicatively* the approximate value is very close to the true value. In contrast, *additively* the two values are *not* close; their difference grows with N and becomes very large. But in virtually all applications of Stirling's formula, it is the multiplicative, as opposed to the additive, comparison that matters.

2. Stirling's formula allows us to answer the question: What is the geometric mean of the first N integers? That is, what is $(N!)^{1/N}$? Using Eq. (3.279), we obtain

$$(N!)^{1/N} \approx \left(N^N e^{-N} \sqrt{2\pi N}\right)^{1/N} = \frac{N}{e}(2\pi N)^{1/2N} \approx \frac{N}{e}, \qquad (3.281)$$

because $N^{1/N} \to 1$ for large N. (The log of $N^{1/N}$ equals $(\ln N)/N$, which goes to zero for large N.) So in a multiplicative sense, the "average" of the first N integers in N/e. A fine result indeed! ♣

(b) The calculation of the higher-order corrections is a bit messier, because we have to keep track of more terms in the Taylor expansion of $f(x)$. To find the order-$1/N$ correction, our strategy will be to write the integrand in Eq. (3.275) as a Gaussian (namely $e^{-y^2/2N}$) plus small corrections. Computing the first four derivatives of $f(x)$, evaluated at N, gives (as you can show) the following modification of Eq. (3.278) and the first line in Eq. (3.279) (letting $y \equiv x - N$, and letting the limits of integration run from $-\infty$ to ∞):

$$N! \approx \int_{-\infty}^{\infty} \exp\left(N \ln N - N - \frac{y^2}{2N} + \frac{y^3}{3N^2} - \frac{y^4}{4N^3}\right) dy$$

$$= N^N e^{-N} \int_{-\infty}^{\infty} \exp\left(-\frac{y^2}{2N}\right) \exp\left(\frac{y^3}{3N^2} - \frac{y^4}{4N^3}\right) dy$$

$$= N^N e^{-N} \int_{-\infty}^{\infty} \exp\left(-\frac{y^2}{2N}\right) \left(1 + \left[\frac{y^3}{3N^2} - \frac{y^4}{4N^3}\right]\right. \qquad (3.282)$$

$$\left. + \frac{1}{2!}\left[\frac{y^3}{3N^2} - \frac{y^4}{4N^3}\right]^2 + \cdots\right) dy,$$

where we have used the Taylor series $e^z \approx 1 + z + z^2/2!$. Since terms with odd powers of y integrate to zero, we obtain (to leading orders in $1/N$),

$$N! \approx N^N e^{-N} \int_{-\infty}^{\infty} \exp\left(-\frac{y^2}{2N}\right)\left(1 - \frac{y^4}{4N^3} + \frac{1}{2}\left[\frac{y^3}{3N^2}\right]^2 + \cdots\right) dy.$$

$$(3.283)$$

At this point, we need to know how to calculate integrals of the form $\int_{-\infty}^{\infty} x^{2n} e^{-ax^2} dx$. Using $\int_{-\infty}^{\infty} e^{-ax^2} dx = \sqrt{\pi} a^{-1/2}$, and successively differentiating both sides with respect to a, we obtain[24]

$$\int_{-\infty}^{\infty} e^{-ax^2}\, dx = \sqrt{\pi}\, a^{-1/2},$$

$$\int_{-\infty}^{\infty} x^2 e^{-ax^2}\, dx = \frac{1}{2}\sqrt{\pi}\, a^{-3/2},$$

$$\int_{-\infty}^{\infty} x^4 e^{-ax^2}\, dx = \frac{3}{4}\sqrt{\pi}\, a^{-5/2},$$

$$\int_{-\infty}^{\infty} x^6 e^{-ax^2}\, dx = \frac{15}{8}\sqrt{\pi}\, a^{-7/2}. \qquad (3.284)$$

Letting $a \equiv 1/2N$ here, Eq. (3.283) gives

$$N! \approx N^N e^{-N} \sqrt{\pi}\left((2N)^{1/2} - \frac{1}{4N^3}\frac{3}{4}(2N)^{5/2} + \frac{1}{18N^4}\frac{15}{8}(2N)^{7/2}\right)$$

$$= N^N e^{-N} \sqrt{2\pi N}\left(1 + \frac{1}{12N}\right). \qquad (3.285)$$

Note that to obtain all the terms of order $1/N$, it is necessary to include the $(y^3/3N^2)^2$ term in Eq. (3.283). This is an easy term to forget.

REMARK: If you like these sorts of calculations, you can go a step further and find the order-$1/N^2$ correction. It turns out that you need to keep terms out to the $-y^6/6N^5$ term in the expansion of $f(x)$ in the first line of Eq. (3.282). Furthermore, you must keep terms out to the $[\cdots]^4/4!$ term in the expansion of e^z in the last line of Eq. (3.282). You can show that the relevant extra terms that take the place of the "$\cdots$" in Eq. (3.283) are then (keeping only terms with even powers of y)

$$\left[-\frac{y^6}{6N^5}\right] + \frac{1}{2!}\left[\left(-\frac{y^4}{4N^3}\right)^2 + 2\left(\frac{y^3}{3N^2}\right)\left(\frac{y^5}{5N^4}\right)\right]$$

$$+ \frac{1}{3!}\left[3\left(\frac{y^3}{3N^2}\right)^2\left(-\frac{y^4}{4N^3}\right)\right] + \frac{1}{4!}\left[\left(\frac{y^3}{3N^2}\right)^4\right], \qquad (3.286)$$

where we have grouped these terms via square brackets according to which term in the e^z series expansion in the last line of Eq. (3.282) they come from. To do all of the necessary integrals in the modified Eq. (3.283), we'll need the next three integrals in the list in Eq. (3.284). They are

$$\int_{-\infty}^{\infty} x^8 e^{-ax^2}\, dx = \frac{3\cdot 5\cdot 7}{2^4}\sqrt{\pi}\, a^{-9/2},$$

$$\int_{-\infty}^{\infty} x^{10} e^{-ax^2}\, dx = \frac{3\cdot 5\cdot 7\cdot 9}{2^5}\sqrt{\pi}\, a^{-11/2},$$

$$\int_{-\infty}^{\infty} x^{12} e^{-ax^2}\, dx = \frac{3\cdot 5\cdot 7\cdot 9\cdot 11}{2^6}\sqrt{\pi}\, a^{-13/2}. \qquad (3.287)$$

---

[24] Yes, it's legal to do this differentiation inside the integral. Integrals are just sums, and differentiating a sum by differentiating each term in it is certainly legal.

Putting the terms of Eq. (3.286) in place of the "···" in Eq. (3.283), you can show that they generate a term equal to $N^N e^{-N} \sqrt{2\pi N}$ times $1/N^2$ times

$$-\frac{1}{6}\frac{3\cdot 5}{2^3}2^3 + \frac{1}{2!}\left(\frac{1}{16}+\frac{2}{15}\right)\frac{3\cdot 5\cdot 7}{2^4}2^4 - \frac{1}{3!}\frac{3}{36}\frac{3\cdot 5\cdot 7\cdot 9}{2^5}2^5$$
$$+\frac{1}{4!}\frac{1}{81}\frac{3\cdot 5\cdot 7\cdot 9\cdot 11}{2^6}2^6$$
$$=\frac{1}{288}.\qquad(3.288)$$

Therefore, we may write Stirling's formula as

$$N! \approx N^N e^{-N} \sqrt{2\pi N}\left(1+\frac{1}{12N}+\frac{1}{288N^2}\right).\qquad(3.289)$$

This result of $1/288$ is rather fortuitous, because it is the third term in the Taylor series for $e^{1/12}$. This means that we can write $N!$ as (with O shorthand for "of order")

$$N! = N^N e^{-N}\sqrt{2\pi N}\left(e^{1/12N}+O(1/N^3)\right)$$
$$\approx N^N e^{-N+1/12N}\sqrt{2\pi N}.\qquad(3.290)$$

It turns out that the order-$1/N^3$ correction is *not* equal to $1/(3!\cdot 12^3)$, which is the next term in the expansion for $e^{1/12}$.

As an example of the increasing accuracy of the various versions of Stirling's formula, let's pick $N = 10$, so $N! = 3{,}628{,}800$. The various Stirling approximations are:

- Eq. (3.279) gives $N! \approx 3{,}598{,}696$. The error is about 1%, consistent with the fact that we haven't included the $1/12N$ term, which is ≈ 0.01.
- Eq. (3.285) gives $N! \approx 3{,}628{,}685$. The error is about 0.003%, consistent with the fact that we haven't included the $1/288N^2$ term, which is ≈ 0.00003.
- Eq. (3.289) and Eq. (3.290) both give $N! \approx 3{,}628{,}810$. The error is about 0.0003%. ♣

53. A handy formula

We'll derive Eqs. (1.5) and (1.6) by deriving the general formula of which they are special cases. As suggested, we'll start with the expression for the sum of an infinite geometric series,

$$1 - a + a^2 - a^3 + a^4 - \cdots = \frac{1}{1+a}.\qquad(3.291)$$

This is valid for $|a| < 1$. (If you plug in, say, $a = 2$, you will get an obviously incorrect statement.) For $|a| < 1$, if you keep enough terms on the left, the sum will be essentially equal to $1/(1+a)$. If you hypothetically keep an *infinite* number of terms, the sum will be *exactly* equal to $1/(1+a)$. You can verify

Eq. (3.291) by multiplying both sides by $1 + a$. On the lefthand side, the infinite number of cross terms cancel in pairs, so only the "1" survives. If we integrate both sides of Eq. (3.291) with respect to a, we obtain

$$a - \frac{a^2}{2} + \frac{a^3}{3} - \frac{a^4}{4} + \frac{a^5}{5} - \cdots = \ln(1 + a). \tag{3.292}$$

Technically there could be a constant of integration in Eq. (3.292), but it is zero (since $a = 0$ correctly yields $0 = \ln(1)$). Eq. (3.292) is the Taylor series for $\ln(1 + a)$. This Taylor series can also be derived via the standard method of taking successive derivatives; see the appendix for a review of Taylor series. As with Eq. (3.291), the result in Eq. (3.292) is valid for $|a| < 1$.

If we now exponentiate both sides of Eq. (3.292), then since $e^{\ln(1+a)} = 1 + a$, we obtain (reversing the sides of the equation)

$$1 + a = e^a e^{-a^2/2} e^{a^3/3} e^{-a^4/4} e^{a^5/5} \cdots, \tag{3.293}$$

which again is valid for $|a| < 1$. Finally, if we raise both sides of Eq. (3.293) to the nth power, we arrive at

$$(1 + a)^n = e^{na} e^{-na^2/2} e^{na^3/3} e^{-na^4/4} e^{na^5/5} \cdots. \tag{3.294}$$

This relation is valid for $|a| < 1$. It is exact if we include an infinite number of the exponential factors on the righthand side. However, the question we are concerned with here is how many terms we need to keep in order to obtain a good approximation. (We'll leave "good" undefined for the moment.) Under what conditions do we obtain Eq. (1.5) or Eq. (1.6)? The number of terms we need to keep depends on both a and n. In the following cases, we will always assume that a is small (more precisely, much smaller than 1).

- $na \ll 1$

    If $na \ll 1$, then all of the exponents on the righthand side of Eq. (3.294) are much smaller than 1. The first one (namely na) is small, by assumption. The second one (namely $na^2/2$; we'll ignore the sign) is also small, because it is smaller than na by a factor a (and also by a factor $1/2$), and we are assuming that a is small. Likewise, all of the other exponents in subsequent terms have additional factors of a and hence are even smaller. Therefore, since all of the exponents in Eq. (3.294) are much smaller than 1 (and since they go to zero quickly enough), they are, to a good approximation, all equal to zero. The exponential factors are therefore all approximately equal to $e^0 = 1$, so we obtain

$$(1 + a)^n \approx 1 \qquad \text{(valid if } na \ll 1) \tag{3.295}$$

    An example of a pair of numbers that satisfies $na \ll 1$ is $n = 1$ and $a = 1/100$. In this case it is a good approximation to say that $(1 + a)^n \approx 1$. And indeed, the exact value of $(1 + a)^n$ is $(1.01)^1 = 1.01$, so the approximation is smaller by only 1%.

- $na^2 \ll 1$

    What if a isn't small enough to satisfy $na \ll 1$, but is still small enough to satisfy $na^2 \ll 1$? In this case we need to keep the e^{na} term in Eq. (3.294), but we can ignore the $e^{-na^2/2}$ term, because it is approximately equal to $e^{-0} = 1$. The exponents in subsequent terms are all also essentially equal to zero, because they are suppressed by higher powers of a. So Eq. (3.294) becomes

    $$(1+a)^n \approx e^{na} \quad \text{(valid if } na^2 \ll 1) \quad (3.296)$$

    We have therefore derived Eq. (1.5), which we now see is valid when $na^2 \ll 1$. A pair of numbers that doesn't satisfy $na \ll 1$ but does satisfy $na^2 \ll 1$ is $n = 100$ and $a = 1/100$. In this case it is a good approximation to say that $(1+a)^n \approx e^{na} = e^1 = 2.718$. And indeed, the exact value of $(1+a)^n$ is $(1.01)^{100} \approx 2.705$, so the approximation is larger by only about 0.5%. The $(1+a)^n \approx 1$ approximation in Eq. (3.295) is not a good one, being smaller than the approximation in Eq. (3.296) by a factor of e in the present scenario.

    A special case of Eq. (3.296) occurs when $n = 1$, which yields $1 + a \approx e^a$. (The lefthand side here is the beginning of the Taylor series for e^a.) Another special case occurs when n is large and $a = 1/n$. (This satisfies $na^2 \ll 1$ since $na^2 = n(1/n)^2 = 1/n$, which is small since n is assumed to be large.) Eq. (3.296) then gives $(1 + 1/n)^n \approx e^1$ (as we saw above with $n = 100$ and $a = 1/100$). This approximation becomes exact in the $n \to \infty$ limit. And indeed, the $n \to \infty$ limit of $(1 + 1/n)^n$ is one way of *defining* the number e.

- $na^3 \ll 1$

    What if a isn't small enough to satisfy $na^2 \ll 1$, but is still small enough to satisfy $na^3 \ll 1$? In this case we need to keep the $e^{-na^2/2}$ term in Eq. (3.294), but we can ignore the $e^{na^3/3}$ term, because it is approximately equal to $e^0 = 1$. The exponents in subsequent terms are all also essentially equal to zero, because they are suppressed by higher powers of a. So Eq. (3.294) becomes

    $$(1+a)^n \approx e^{na} e^{-na^2/2} \quad \text{(valid if } na^3 \ll 1) \quad (3.297)$$

    We have therefore derived Eq. (1.6), which we now see is valid when $na^3 \ll 1$. A pair of numbers that doesn't satisfy $na^2 \ll 1$ but does satisfy $na^3 \ll 1$ is $n = 10{,}000$ and $a = 1/100$. In this case it is a good approximation to say that $(1+a)^n \approx e^{na} e^{-na^2/2} = e^{100} e^{-1/2} = 1.6304 \cdot 10^{43}$. And indeed, the exact value of $(1+a)^n$ is $(1.01)^{10{,}000} \approx 1.6358 \cdot 10^{43}$, so the approximation is smaller by only about 0.3%. The $(1+a)^n \approx e^{na}$ approximation in Eq. (3.296) is not a good one, being larger than the approximation in Eq. (3.297) by a factor of $e^{1/2}$ in the present scenario.

We can continue in this manner. If a isn't small enough to satisfy $na^3 \ll 1$, but is still small enough to satisfy $na^4 \ll 1$, then we need to keep the $e^{na^3/3}$ term in Eq. (3.294), but we can set the $e^{-na^4/4}$ term (and all subsequent terms) equal to 1. And so on and so forth. However, it is rare that you will need to go beyond the two terms in Eq. (3.297). Theoretically though, if, say, $n = 10^{12}$ and $a = 1/100$, then we need to keep the terms in Eq. (3.294) out to the $e^{-na^6/6}$ term, but we can ignore the $e^{na^7/7}$ term and beyond, to a good approximation.

In any case, the rough size of the (multiplicative) error is the first term in Eq. (3.294) that is dropped. This is true because however close the first dropped term is to $e^0 = 1$, all of the subsequent exponential factors are even closer to $e^0 = 1$. In the $n = 10,000$ and $a = 1/100$ case in the third bullet point above, the multiplicative error is roughly equal to the $e^{na^3/3}$ factor that we dropped, which in this case equals $e^{1/300} \approx 1.0033$. This is approximately the factor by which the true answer is larger than the approximate one.[25] This agrees with the results we found above, because $(1.6358)/(1.6304) \approx (1.0033)$. The true answer is larger by about 0.3% (so the approximation is smaller by about 0.3%).

If this factor of 1.0033 is close enough to 1 for whatever purpose we have in mind, then the approximation is a good one. If it isn't close enough to 1, then we need to keep additional terms until it is. In the present example with $n = 10,000$ and $a = 1/100$, if we keep the $e^{na^3/3}$ factor, then the multiplicative error is essentially equal to the next term in Eq. (3.294), which is $e^{-na^4/4} = e^{-1/40,000} = 0.999975$. This is approximately the factor by which the true answer is smaller than the approximate one. The difference is only 0.0025%.

54. **Exponential distribution**

Let's first quickly verify that the rate λ is indeed given by $\lambda = 1/\tau$, where τ is the average time between events. Consider a large time t. From the definition of τ, the expected number of events that occur during the time t is t/τ. But also, from the definition of λ (the number of events per second), another expression for the expected number of events is λt. Equating these two expressions gives $\lambda = 1/\tau$, as desired. The rate is therefore simply the reciprocal of the average waiting time. For example, if the waiting time is $1/5$ of a second, then the rate is 5 events per second, which makes sense.

We'll now determine the probability distribution $\rho(t)$ of the waiting time to the next event. That is, we'll determine the probability $\rho(t)\,dt$ that the waiting time to the next event is between t and $t + dt$, where dt is small. To do this, we'll divide time into very small intervals with length ϵ. We'll then take the $\epsilon \to 0$ limit, which is equivalent to making time be continuous.

The division of time into little intervals is summarized in Fig. 3.53. Time 0 is when we start our stopwatch and begin waiting for the next event. (An event need not actually occur at $t = 0$.) From time 0 to time t, there are t/ϵ (which is a very large number) of intervals, each with the very small length ϵ. And then the dt interval sits at the end. Both ϵ and dt are assumed to be very small, but they need not have anything to do with each other. ϵ exists as a calculational tool only,

---

[25]The exponent here is positive, which means that the factor is slightly larger than 1. But note that half of the terms in Eq. (3.294) have negative exponents. If one of those terms is the first one that is dropped, then the factor is slightly smaller than 1. This is approximately the factor by which the true answer is smaller than the approximate one.

while dt is the arbitrarily chosen small time interval that appears in the $\rho(t)\,dt$ probability we are trying to find.

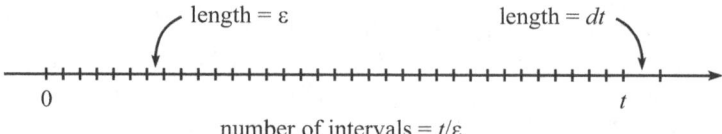

Figure 3.53

In order for the next success (event) to happen between t and $t + dt$, there must be failure during every one of the t/ϵ intervals of length ϵ shown in Fig. 3.53, and then there must be success between t and $t + dt$.

REMARK: To proceed, we'll need to know the probability that success happens in a small time interval with length dt (or ϵ). We claim that this probability is $\lambda\,dt$. In other words, not only is $\lambda\,dt$ the *expected* number of events in the time dt, it is also essentially equal to the *probability* that an event (that is, at least one event) occurs in the time dt. This is true because dt is assumed to be infinitesimal. (It certainly isn't true for large dt, because if dt is large enough, then $\lambda\,dt$ is greater than 1, so it can't represent a probability.) The reason why $\lambda\,dt$ is the probability when dt is infinitesimal is because the probability of one event occurring is so small that we don't need to worry about multiple events occurring. We can be explicit about this as follows. We know that the expected number of events during any arbitrary time T is λT. But another expression for the expected number of events is the sum of each number weighted by the probability of that number of events occurring. Therefore,

$$\lambda T = P_T(0) \cdot 0 + P_T(1) \cdot 1 + P_T(2) \cdot 2 + P_T(3) \cdot 3 + \cdots, \qquad (3.298)$$

where $P_T(k)$ is the probability of obtaining exactly k events during the time T. Solving for $P_T(1)$ gives

$$P_T(1) = \lambda T - P_T(2) \cdot 2 - P_T(3) \cdot 3 - \cdots. \qquad (3.299)$$

We see that $P_T(1)$ is smaller than λT due to the $P_T(2)$ and $P_T(3)$, etc., probabilities. So $P_T(1)$ isn't equal to λT. However, if all of the probabilities of multiple events occurring ($P_T(2)$, $P_T(3)$, etc.) are very small, then $P_T(1)$ is *essentially* equal to λT. And this is exactly what happens if the time interval T is very small, that is, if T is replaced by dt. For small time intervals dt, there is hardly any chance of the event even occurring *once*. So it is even less likely that it will occur *twice*, and even less likely for three times, etc., because these scenarios involve higher powers of a small probability. We therefore see that the probability of at least one event occurring during the time dt is essentially equal to the probability of exactly one event occurring, which in turn is essentially equal to $\lambda\,dt$. ♣

Returning to the problem, we can say that since ϵ is infinitesimal, the probability of success in any given small interval of length ϵ is $\lambda\epsilon$, which means that the probability of failure is $1 - \lambda\epsilon$. And since there are t/ϵ of these intervals, the probability of failure in all of them is $(1 - \lambda\epsilon)^{t/\epsilon}$. The probability that the next success (that is, the first one after $t = 0$) happens between t and $t + dt$ (which is $\rho(t)\,dt$ by the definition of the probability density $\rho(t)$) equals the probability of failure during every one of the t/ϵ intervals of length ϵ, multiplied by the probability of success between t and $t + dt$. Therefore,

$$\rho(t)\,dt = \left((1 - \lambda\epsilon)^{t/\epsilon}\right)(\lambda\,dt). \tag{3.300}$$

We'll now invoke the first result from Problem 53, namely

$$(1 + a)^n \approx e^{na}. \tag{3.301}$$

This holds for negative a as well as positive a. For the case at hand, a comparison of Eqs. (3.300) and (3.301) shows that we want to define $a \equiv -\lambda\epsilon$ and $n \equiv t/\epsilon$, which yields $na = -\lambda t$. Eq. (3.301) then gives $(1 - \lambda\epsilon)^{t/\epsilon} \approx e^{-\lambda t}$, so Eq. (3.300) yields

$$\rho(t) = \lambda e^{-\lambda t}. \tag{3.302}$$

This is the desired probability distribution of the waiting time to the next event. It is called the "exponential distribution" since the distribution decreases exponentially with t. If you want to work in terms of the average waiting time τ instead of the rate λ, the preceding result becomes (using $\lambda = 1/\tau$)

$$\rho(t) = \frac{e^{-t/\tau}}{\tau}. \tag{3.303}$$

Note that whichever way we choose to write it, the exponential distribution is completely specified by just one parameter, either λ or τ.

Fig. 3.54 shows plots of $\rho(t)$ for a few different values of the average waiting time, τ. The two main properties of each of these curves are the starting value at $t = 0$ and the rate of decay as t increases. From Eq. (3.303), the starting value at $t = 0$ is $e^0/\tau = 1/\tau$. So the bigger τ is, the smaller the starting value. This makes sense, because if the average waiting time τ is large (equivalently, if the rate λ is small), then there is only a small chance that the next event will happen right away.

How fast do the curves decay? This is governed by the denominator of the exponent in Eq. (3.303). For every τ units that t increases by, $\rho(t)$ decreases by a factor of $e^{-\tau/\tau} = 1/e$. If τ is large, the curve takes longer to decrease by a factor of $1/e$. This is consistent with Fig. 3.54, where the large-τ curve falls off slowly, and the small-τ curve falls off quickly. To sum up, if τ is large, the $\rho(t)$ curve starts off low and decays slowly. And if τ is small, the curve starts off high and decays quickly.

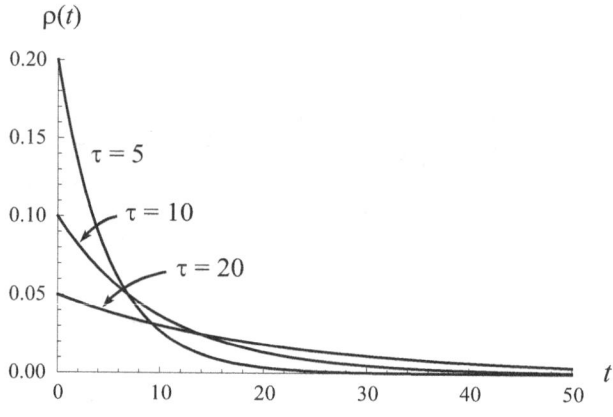

Figure 3.54

REMARKS:

1. $\rho(t)$ is often referred to as the probability distribution of the waiting time from one event to the next. While this is true, a more general statement holds: $\rho(t)$ is the probability distribution of the waiting time from *any point in time* to the occurrence of the next event. That is, you can start your stopwatch at any time, not just at the occurrence of an event. This is consistent with the wording in the statement of the problem. We didn't specify that an event occurred at the starting time. And if you look back through the above solution, you will see that nowhere did we assume that an event actually occurred at $t = 0$.

   However, beware of the following incorrect reasoning. Let's say that an event happens at $t = 0$, but that you don't start your stopwatch until, say, $t = 1$. The fact that the next event after $t = 1$ doesn't happen (on average) until $t = 1 + \tau$ (from the previous paragraph) seems to imply that the average waiting time starting at $t = 0$ is $1 + \tau$. But it better not be, because we know from above that it's just τ. The error here is that we forgot about the scenarios where the next event after $t = 0$ happens *between* $t = 0$ and $t = 1$. When these events are included, the average waiting time, starting at $t = 0$, ends up correctly being τ. (As an exercise, you can verify this by considering separately the cases where the next event happens before $t = 1$ or after $t = 1$.) In short, the waiting time from $t = 1$ is indeed τ, but the next event (after the $t = 0$ event) might have already happened before $t = 1$.

2. The waiting time has to be *something*, so the sum of the $\rho(t)\,dt$ probabilities, over all the possible values of t, must be 1. The sum of these probabilities is just the integral $\int_0^\infty \rho(t)\,dt$. You can quickly verify that this equals 1. In other words, the area under each of the curves in Fig. 3.54 is 1.

   Likewise, the expectation value (the average value) of the waiting time between events must be τ, because that is how τ was defined. The expectation value is the sum of the t values, weighted by the $\rho(t)\,dt$ probabilities. So the expectation value is the integral $\int_0^\infty t \cdot \rho(t)\, dt$. You can verify that this equals τ.

3. In a sense, the curves for all of the different values of τ in Fig. 3.54 are really the same curve. They're just stretched or squashed in the horizontal and vertical directions. The general form of the curve described by the expression in Eq. (3.303) is shown in Fig. 3.55.

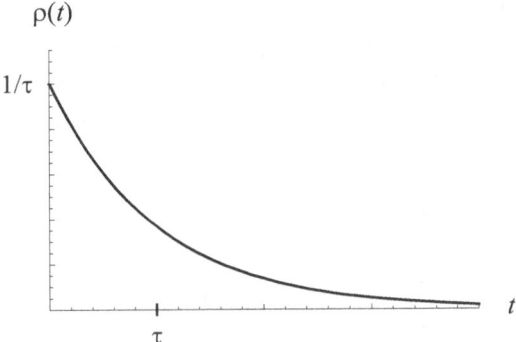

Figure 3.55

As long as we change the scales on the axes so that τ and $1/\tau$ are always located at the same positions, then the curves will look the same for any τ. For example, as we noted above, no matter what the value of τ is, the value of the curve at $t = \tau$ is always $1/e$ times the value at $t = 0$. Of course, when we plot things, we usually keep the scales fixed, in which case the τ and $1/\tau$ positions move along the axes, as shown in Fig. 3.56 (these are the same curves as in Fig. 3.54). But by suitable uniform stretching/squashing of the axes, each of these curves can be turned into the curve in Fig. 3.55 (and vice versa).

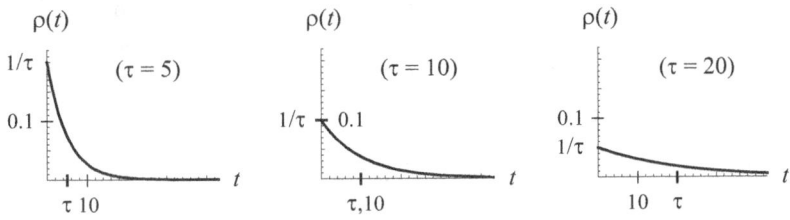

Figure 3.56

The fact that any of the curves in Fig. 3.56 can be obtained from any of the other curves by stretching and squashing the two directions by inverse (as you can verify) factors implies that the areas under all of the curves are the same. This is consistent with the fact that all of the areas must be 1 (since the total probability must be 1).

4. We phrased the exponential distribution in terms of waiting *times*, but the distribution also applies to waiting *distances*, or any other parameter for which the events happen completely randomly. For example, if we assume that typos occur at random locations in a book, and if we know the average distance τ between them (measured in pages, not necessarily integral), then Eq. (3.303) gives the distribution of waiting distances (1) between typos, and (2) from any random point to the next typo. ♣

55. Poisson distribution

As in the solution to Problem 54, the random process can be completely described by just *one* number – the average rate of events, which we'll again call λ. As we saw in Problem 54, $\lambda\epsilon$ is the probability that exactly one event occurs in a very small time interval ϵ.

Our goal here is to answer the question: What is the probability, $P(k)$, that exactly k events occur during a given time span of t? To answer this, we'll divide time into very small intervals with length ϵ. We'll then take the $\epsilon \to 0$ limit, which is equivalent to making time be continuous. The division of time into little intervals is summarized in Fig. 3.57. There are t/ϵ intervals, which we'll label as n.

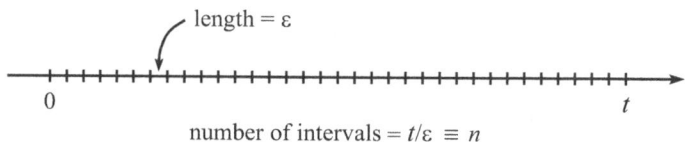

Figure 3.57

The probability that a *specific set* of k of the n little intervals all yield exactly one event each equals $(\lambda\epsilon)^k$, because each of the k intervals has a $\lambda\epsilon$ probability of yielding one event. We then need the other $n - k$ intervals to *not* yield an event, because we want *exactly* k events. This happens with probability $(1 - \lambda\epsilon)^{n-k}$, because each of the $n - k$ intervals has a $1 - \lambda\epsilon$ chance of yielding zero events. The probability that a specific set of k intervals (and no others) each yield an event is therefore $(\lambda\epsilon)^k \cdot (1 - \lambda\epsilon)^{n-k}$. Finally, since there are $\binom{n}{k}$ ways to pick a specific set of k intervals, we see that the probability that exactly k of the n intervals each yield an event is

$$P(k) = \binom{n}{k}(\lambda\epsilon)^k(1 - \lambda\epsilon)^{n-k}. \tag{3.304}$$

This is simply the standard binomial distribution with the usual probability p replaced with $\lambda\epsilon$.

Let's now see what Eq. (3.304) reduces to in the $\epsilon \to 0$ limit. Note that $\epsilon \to 0$ implies $n \equiv t/\epsilon \to \infty$. If we write out the binomial coefficient and expand things a bit, Eq. (3.304) becomes

$$P(k) = \frac{n!}{(n-k)!\,k!}(\lambda\epsilon)^k(1 - \lambda\epsilon)^n(1 - \lambda\epsilon)^{-k}. \tag{3.305}$$

Of the various letters in this equation, n is huge, ϵ is tiny, and λ and k are "normal," not assumed to be huge or tiny. λ is determined by the setup, and k is the number of events we're concerned with. (We'll see below that the relevant k's are roughly the size of the expected number of events in the time t, which is $\lambda t = \lambda n \epsilon$.) In the $\epsilon \to 0$ limit (and hence $n \to \infty$ limit), we can make three simplifications to Eq. (3.305):

- First, in the $n \to \infty$ limit, we can say that[26]

$$\frac{n!}{(n-k)!} = n^k, \quad (3.306)$$

at least in a multiplicative sense (we don't care about an additive sense). This follows from the fact that $n!/(n-k)!$ is the product of the k numbers from n down to $n - k + 1$. And if n is large compared with k (which it certainly is, in the $n \to \infty$ limit), then all of these k numbers are essentially (or exactly, in the $n \to \infty$ limit) equal to n (multiplicatively). Therefore, since there are k of them, we obtain n^k.

- Second, we can use the $(1 + a)^n \approx e^{na}$ approximation from Eq. (1.5) in Problem 53 (which is exact in the $\epsilon \to 0$ and $n \to \infty$ limits) to simplify the $(1 - \lambda \epsilon)^n$ term. With $a \equiv -\lambda \epsilon$, Eq. (1.5) gives

$$(1 - \lambda \epsilon)^n = e^{-n\lambda \epsilon}. \quad (3.307)$$

- Third, in the $\epsilon \to 0$ limit, we can use the $(1 + a)^n \approx e^{na}$ approximation again, this time to simplify the $(1 - \lambda \epsilon)^{-k}$ term. The result is

$$(1 - \lambda \epsilon)^{-k} = e^{k\lambda \epsilon} = e^0 = 1, \quad (3.308)$$

because for any fixed values of k and λ, the $k\lambda \epsilon$ exponent becomes infinitesimally small as $\epsilon \to 0$. Basically, in $(1 - \lambda \epsilon)^{-k}$ we're forming a finite power of a number that is essentially equal to 1. Note that this reasoning doesn't apply to the $(1 - \lambda \epsilon)^n$ term in Eq. (3.307), because n isn't a fixed number. It changes with ϵ, in that it becomes large when ϵ becomes small.

Applying these three simplifications to Eq. (3.305) gives

$$\begin{aligned} P(k) &= \frac{n^k}{k!}(\lambda \epsilon)^k e^{-n\lambda \epsilon} \cdot 1 \\ &= \frac{1}{k!}(\lambda \cdot n\epsilon)^k e^{-\lambda \cdot n\epsilon} \\ &= \frac{1}{k!}(\lambda t)^k e^{-\lambda t}, \end{aligned} \quad (3.309)$$

where we have used $n \equiv t/\epsilon \implies n\epsilon = t$. Now, from the definition of the rate λ, λt is the average (expected) number of events that occur in the time t. Let's label

---

[26] All three simplifications here would involve "$\approx$" signs if we were simply dealing with large values of n. But since we're actually taking the $n \to \infty$ limit, the "$\approx$" signs become "=" signs.

this average number of events as $a \equiv \lambda t$. We can then write Eq. (3.309) as

$$P(k) = \frac{a^k e^{-a}}{k!}, \qquad (3.310)$$

where a is the average number of events in the time interval under consideration. This is the desired Poisson distribution. It gives the probability of obtaining exactly k events during a period of time for which the average number is a. Note that while the *observed* number of events k must be an integer, the *average* number of events a need not be.

REMARKS:

1. Since a is the only parameter left on the righthand side of Eq. (3.310), the distribution is completely specified by a. The individual values of λ and t don't matter. All that matters is their product $a \equiv \lambda t$. This means that if we, say, double the time interval t under consideration and also cut the rate λ in half, then a remains unchanged. So we have exactly the same distribution $P(k)$. Although it is clear that doubling t and halving λ yields the same *average* number of events (since the average equals the product λt), it might not be intuitively obvious that the entire $P(k)$ *distribution* is the same. But the result in Eq. (3.310) shows that this is indeed the case.

2. The Poisson distribution in Eq. (3.310) works perfectly well for small a, even $a < 1$. It's just that in this case, the plot of $P(k)$ doesn't have a bump in it. Instead, it starts high and then falls off as k increases. Fig. 3.58 shows the plot of $P(k)$ for various values of a. We've arbitrarily decided to cut off the plots at $k = 20$, even though they technically go on forever. We can theoretically have an arbitrarily large number of events in any given time interval, although the probability is negligibly small. In the plots, the probabilities are effectively zero by $k = 20$, except in the $a = 15$ case.

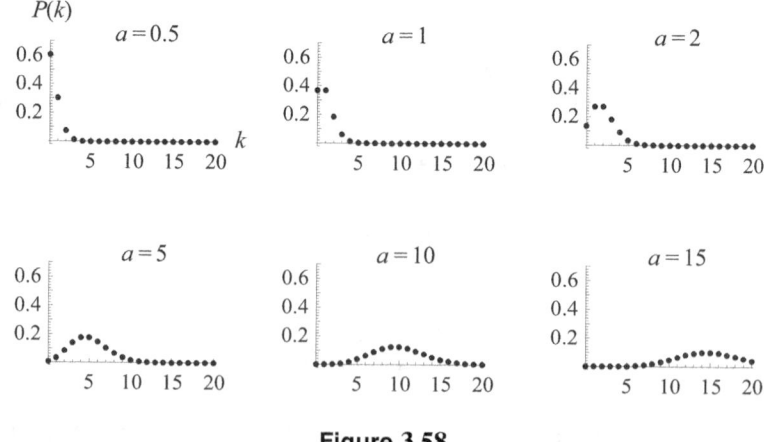

Figure 3.58

As a increases, the bump in the plots (once it actually becomes a bump) does three things (as you can show): (1) it shifts to the right (it is centered near $k = a$), (2) it decreases in height, and (3) it becomes wider. The last two of these properties are consistent with each other, in view of the fact that the sum of all the probabilities must equal 1, for any value of a.

Eq. (3.310) gives the probability of obtaining zero events as $P(0) = e^{-a}$. If $a = 0.5$ then $P(0) = e^{-0.5} \approx 0.61$. This agrees with a visual inspection of the first plot in Fig. 3.58. Likewise, if $a = 1$ then $P(0) = e^{-1} \approx 0.37$ (and $P(1)$ takes on this same value), in agreement with the second plot. If a is large then the $P(0) = e^{-a}$ probability goes to zero, in agreement with the bottom three plots. This makes sense; if the average number of events is *large*, then it is very *unlikely* that we obtain zero events. In the opposite extreme, if a is very small (for example, $a = 0.01$), then the $P(0) = e^{-a}$ probability is very close to 1. This again makes sense; if the average number of events is very *small*, then it is very *likely* that we obtain zero events.

To make it easier to compare the six plots in Fig. 3.58, we have superimposed them in Fig. 3.59. Although we have drawn these Poisson distributions as continuous curves to make things clearer, remember that the distribution applies only to integer values of k.

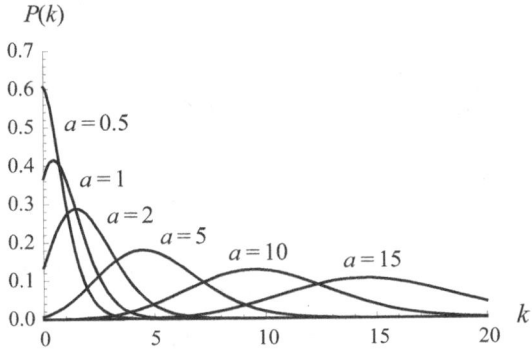

Figure 3.59

3. You are encouraged to investigate various aspects of the Poisson distribution, namely: the fact that the total probability is 1, the location of the maximum, an approximate value of the maximum when a is large, the expectation value, and the standard deviation. (The standard deviation is defined in the 7th remark in the solution to Problem 56.)

Some answers so you can check your work: The maximum is located at the integer value of k that lies between $a - 1$ and a (or at both of these values if a is an integer; this is consistent with the plots in Fig. 3.58). When a is large, the value of the maximum is approximately $1/\sqrt{2\pi a}$. The expectation value is a (of course, since that's how a was defined). And the standard deviation is $\sqrt{a}$. ♣

56. Gaussian approximation to the binomial distribution

The first step in transforming the binomial expression to the Gaussian one is to use Stirling's approximation, $N! \approx N^N e^{-N} \sqrt{2\pi N}$, to rewrite each of the three factorials in the binomial coefficient in Eq. (1.10). This gives

$$\binom{2n}{n+x} = \frac{(2n)!}{(n+x)!(n-x)!} \tag{3.311}$$

$$\approx \frac{(2n)^{2n} e^{-2n} \sqrt{2\pi(2n)}}{\left[(n+x)^{n+x} e^{-(n+x)} \sqrt{2\pi(n+x)}\right] \cdot \left[(n-x)^{n-x} e^{-(n-x)} \sqrt{2\pi(n-x)}\right]}.$$

Canceling all the e's and a few other factors yields

$$\binom{2n}{n+x} \approx \frac{(2n)^{2n} \sqrt{n}}{(n+x)^{n+x}(n-x)^{n-x} \sqrt{\pi} \sqrt{n^2 - x^2}}. \tag{3.312}$$

Let's now divide both the numerator and denominator by n^{2n}. In the denominator, we'll do this by dividing the first and second factors by n^{n+x} and n^{n-x}, respectively. The result is

$$\binom{2n}{n+x} \approx \frac{2^{2n} \sqrt{n}}{\left(1 + \frac{x}{n}\right)^{n+x} \left(1 - \frac{x}{n}\right)^{n-x} \sqrt{\pi} \sqrt{n^2 - x^2}}. \tag{3.313}$$

We'll now apply the $(1+a)^m \approx e^{ma} e^{-ma^2/2}$ approximation from Eq. (1.6) in Problem 53. (And yes, we do need both terms on the righthand side. The simpler approximation in Eq. (1.5) would yield the wrong answer. See the fifth remark below.) With a and m defined to be $a \equiv x/n$ and $m \equiv n+x$, we have (using the notation $\exp(y)$ for e^y, to avoid writing lengthy exponents)

$$\left(1 + \frac{x}{n}\right)^{n+x} \approx \exp\left((n+x)\left(\frac{x}{n}\right) - \frac{1}{2}(n+x)\left(\frac{x}{n}\right)^2\right). \tag{3.314}$$

When we multiply things out here, we find that there is a $-x^3/2n^2$ term. However, we'll see below that the x's we'll be dealing with are much smaller than n, which means that the $-x^3/2n^2$ term is much smaller than the other terms. So we'll ignore it. We are then left with

$$\left(1 + \frac{x}{n}\right)^{n+x} \approx \exp\left(x + \frac{x^2}{2n}\right). \tag{3.315}$$

Although the $x^2/2n$ term here is much smaller than the x term (assuming $x \ll n$), we do in fact need to keep it, because the x term will cancel in Eq. (3.317) below. (The $-x^3/2n^2$ term would actually cancel too, for the same reason.) In a similar manner, we obtain

$$\left(1 - \frac{x}{n}\right)^{n-x} \approx \exp\left(-x + \frac{x^2}{2n}\right). \tag{3.316}$$

Eq. (3.313) therefore becomes

$$\binom{2n}{n+x} \approx \frac{2^{2n} \sqrt{n}}{\exp\left(x + \frac{x^2}{2n}\right) \exp\left(-x + \frac{x^2}{2n}\right) \sqrt{\pi} \sqrt{n^2 - x^2}}. \tag{3.317}$$

When combining (adding) the exponents, the x and $-x$ cancel. Also, under the assumption that $x \ll n$, we can say that $\sqrt{n^2 - x^2} \approx \sqrt{n^2 - 0} = n$. Eq. (3.317) then becomes

$$\binom{2n}{n+x} \approx \frac{2^{2n} \sqrt{n}}{e^{x^2/n} \sqrt{\pi} \, n}. \tag{3.318}$$

Finally, if we substitute Eq. (3.318) into Eq. (1.10), the 2^{2n} factors cancel, and we are left with the desired result,

$$P_B(x) \approx \frac{e^{-x^2/n}}{\sqrt{\pi n}} \equiv P_G(x) \qquad \text{(for $2n$ coin flips)} \tag{3.319}$$

This is the probability of obtaining $n + x$ Heads in $2n$ coin flips. If we want to switch back to having the number of flips be n instead of $2n$, then we just need to replace n with $n/2$ in Eq. (3.319). The result is (with x now being the deviation from $n/2$ Heads)

$$P_B(x) \approx \frac{e^{-2x^2/n}}{\sqrt{\pi n/2}} \equiv P_G(x) \qquad \text{(for n coin flips)} \tag{3.320}$$

Whether you use Eq. (3.319) or Eq. (3.320), the coefficient of π and the inverse of the coefficient of x^2 are both equal to half the number of flips.

If you want to write the above results in terms of the actual number k of Heads, instead of the number x of Heads relative to the expected number, you can just replace x with either $k - n$ in Eq. (3.319), or $k - n/2$ in Eq. (3.320).

Let's see how accurate Eq. (3.319) (or Eq. (3.320)) is. Consider, for example, the probability of getting 45 Heads in 100 flips. The exact probability in Eq. (1.10) (with $n = 50$ and $x = -5$) is $\binom{100}{45}/2^{100} \approx 0.04847$, and the approximate probability in Eq. (3.319) is $e^{-(-5)^2/50}/\sqrt{\pi \cdot 50} \approx 0.04839$. The error is only about 0.17%.

REMARKS:

1. In the above derivation, we claimed that if n is large (as we are assuming), then any values of x that we are concerned with are much smaller than n. This allowed us to simplify various expressions by ignoring certain terms. Let's be explicit about how the logic of the $x \ll n$ assumption proceeds.

   What we showed above (assuming n is large) is that *if* the $x \ll n$ condition is satisfied, *then* Eq. (3.319) is valid. And the fact of the matter is that if n is large, we'll never be interested in values of x that don't satisfy $x \ll n$ (and hence for which Eq. (3.319) might not be valid), because the associated probabilities are negligible. This is true because if, for example, $x = 10\sqrt{n}$ (which certainly satisfies $x \ll n$ if n is large, which means that Eq. (3.319) is indeed valid), then the $e^{-x^2/n}$ exponential factor in Eq. (3.319) equals $e^{-10^2} = e^{-100} \approx 4 \cdot 10^{-44}$, which is completely negligible. (Even if x is only $2\sqrt{n}$, the $e^{-x^2/n}$ factor equals $e^{-2^2} = e^{-4} \approx 0.02$.) Larger values of x will yield even smaller probabilities, because we know that the binomial coefficient in Eq. (1.10) decreases as x gets farther from zero. (This is evident if you look at a typical Pascal's triangle of binomial coefficients.) These probabilities might not satisfy Eq. (3.319), but we don't care, because they're so small.

2. The most important part of the Gaussian distribution is the n in the denominator of the exponent, because this (or rather, its square root) determines the rough width of the distribution. We'll have more to say about this in Remarks 7 and 8 below.

3. Since x appears only through its square, $P_G(x)$ is an even function of x. That is, x and $-x$ yield the same value of the function; it is symmetric around $x = 0$. This evenness makes intuitive sense, because we're just as likely to get, say, four Heads above the average as four Heads below the average.

4. The probability that exactly half (that is, n) of $2n$ coin flips come up Heads is obtained by plugging $x = 0$ into Eq. (3.319). The result is $P_G(0) = e^{-0}/\sqrt{\pi n} = 1/\sqrt{\pi n}$. (You can also obtain this by applying Stirling's formula directly to the binomial probability $\binom{2n}{n}/2^{2n}$.) For example, with 100 flips the probability of obtaining exactly 50 Heads is $1/\sqrt{50\pi} \approx 8\%$.

5. Note that we really did need the $e^{-ma^2/2}$ factor in the approximation from Eq. (1.6). If we had used the less accurate $(1 + a)^m \approx e^{ma}$ version from Eq. (1.5), we would have had incorrect x^2/n terms in Eqs. (3.315) and (3.316), instead of the correct $x^2/2n$ terms.

6. With n flips, the sum of all the $\binom{n}{k}/2^n$ binomial probabilities must equal 1, of course. In other words, $\sum_{k=0}^{n} \binom{n}{k}$ must equal 2^n. This is indeed true, because this sum is what arises when applying the binomial expansion to the lefthand side of $(1 + 1)^n = 2^n$. The sum of the Gaussian probabilities in Eq. (3.319) must likewise equal 1 (at least in the approximation where the Gaussian expression is valid). For large n, we can approximate the sum by an integral, and we can extend the integral to $\pm\infty$ with negligible error (because the probabilities are so small in the extremes). It must therefore be true that

$$\int_{-\infty}^{\infty} \frac{e^{-x^2/n}}{\sqrt{\pi n}} dx = 1. \tag{3.321}$$

This equality does indeed hold, because $\int_{-\infty}^{\infty} e^{-y^2/n} dy = \sqrt{\pi n}$; see Eq. (3.280) in the solution to Problem 52 for a proof.

7. Written in terms of the *standard deviation* σ, the general formula for the Gaussian distribution (with a mean value of zero) is

$$f(x) = \sqrt{\frac{1}{2\pi\sigma^2}} e^{-x^2/2\sigma^2}. \tag{3.322}$$

The standard deviation of a distribution has a formal definition: It is the square root of the average (expected) value of the square of the distance from the mean μ. That is, $\sigma \equiv \sqrt{E[(x-\mu)^2]}$, where the "$E$"

stands for the expected value. In our case where the mean μ is zero, we just have $\sigma = \sqrt{E[x^2]}$. Let's show that the standard deviation of the Gaussian distribution in Eq. (3.322) is indeed the σ that appears in the formula. We need to evaluate the integral $E[x^2] = \int_{-\infty}^{\infty} x^2 f(x)\,dx$ and then take the square root. Eq. (3.284) in the solution to Problem 52 tells us that $\int_{-\infty}^{\infty} x^2 e^{-ax^2} dx = (1/2)\sqrt{\pi} a^{-3/2}$. Letting $a \equiv 1/2\sigma^2$ yields $\int_{-\infty}^{\infty} x^2 e^{-x^2/2\sigma^2} dx = (1/2)\sqrt{\pi}(2\sigma^2)^{3/2} = \sqrt{2\pi}\sigma^3$. Including the $1/\sqrt{2\pi}\sigma$ prefactor in Eq. (3.322) then gives $\int_{-\infty}^{\infty} x^2 f(x) = \sigma^2$. By definition, the standard deviation is the square root of this integral, which gives σ, as desired.

If we compare the Gaussian result in Eq. (3.320) (for n coin flips) with the expression in Eq. (3.322), we see that they agree if $\sigma = \sqrt{n/4}$. This correspondence makes *both* the prefactor and the coefficient of x^2 in the exponent agree. The standard deviation of our Gaussian approximation in Eq. (3.320) is therefore $\sigma = \sqrt{n/4}$. This gives a rough measure of the spread of the Gaussian curve. When $x = \pm\sigma$, we have $f(x) = e^{-1/2} f(0) \approx (0.61) f(0)$. When $x = \pm 3\sigma$, we have $f(x) = e^{-9/2} f(0) \approx (0.01) f(0)$. And when $x = \pm 5\sigma$, we have $f(x) = e^{-25/2} f(0) \approx (4 \cdot 10^{-6}) f(0)$. So $f(x)$ is nearly zero at $x = \pm 3\sigma$, and essentially zero at $x = \pm 5\sigma$.

8. The fact that the standard deviation σ is proportional to $\sqrt{n}$, as opposed to n, has huge implications. Since $\sqrt{n}$ is negligible compared with n when n is large, the *relative* width of the Gaussian bump (compared with the full range of possible values, which is n if you're flipping n coins) is proportional to $\sqrt{n}/n$, which goes to zero for large n. The *fractional* deviation of the number of Heads from the average therefore goes to zero for large n. In other words, if you flip a very large number of coins, you're essentially guaranteed to get pretty much 50% Heads. This is known as the *law of large numbers*. For example, if you flip a million coins, there is no chance that the number of Heads will differ from the average by more than 1% (that is, by more than 10,000). By "no chance" we mean that the probability can be shown to be of order 10^{-88}.

To get an idea of how ridiculously small the number 10^{-88} is, imagine (quite hypothetically, of course) gathering together as many people as there are protons and neutrons in the earth (roughly $4 \cdot 10^{51}$), and imagine each person running the given experiment (flipping a million coins) once a second for the entire age of the universe (roughly $4 \cdot 10^{17}$ seconds). And then repeat this whole process ten quintillion (10^{19}) times. This will yield $1.6 \cdot 10^{88}$ runs of the experiment, in which case you might expect one or two runs to have percentages of Heads that differ from 50% by more than 1%.

9. If the two probabilities involved in a binomial distribution are p and $1-p$ instead of the two $1/2$'s in the case of a coin toss, then the probability of k successes in n trials is $P(k) = \binom{n}{k} p^k (1-p)^{n-k}$. (This is true because $p^k(1-p)^{n-k}$ is the probability that a *particular* set of k trials yield success while the complementary $n-k$ trials yield failure. And there are $\binom{n}{k}$ ways to choose the set of k trials that are successful.) If, for example, we're

concerned with the number of 5's we obtain in n rolls of a die, then $p = 1/6$.
It turns out that for large n, the binomial distribution $P(k)$ is essentially a Gaussian distribution for *any* value of p, not just the $p = 1/2$ value we dealt with above. The Gaussian is centered around the expectation value of k (namely pn), as you would expect. The derivation of this Gaussian form follows the same steps as above. But it gets rather messy, so we'll just state the result: For large n, the probability of obtaining $k = pn + x$ successes in n trials is approximately equal to

$$P_G(x) = \frac{e^{-x^2/[2np(1-p)]}}{\sqrt{2\pi np(1-p)}} \qquad \text{(for n trials with a general p)} \qquad (3.323)$$

If $p = 1/2$, this reduces to the result in Eq. (3.320), as it should.

Eq. (3.323) implies that the bump in the plot of $P_G(x)$ is symmetric around $x = 0$ (or equivalently, around $k = pn$) for *any* p, not just $p = 1/2$. This isn't so obvious, because for $p \neq 1/2$ the bump isn't centered around $n/2$. That is, the *location* of the bump is lopsided with respect to $n/2$. So you might think that the *shape* of the bump should be lopsided too. But it isn't. (Well, the tail extends farther to one side, but $P_G(x)$ is essentially zero in the tails.) Fig. 3.60 shows a plot of Eq. (3.323) for $p = 1/6$ and $n = 60$, which corresponds to rolling a die 60 times and seeing how many, say, 5's you get. The $x = 0$ point corresponds to having $pn = (1/6)(60) = 10$ rolls of a 5. The bump is symmetric (although the true $P_B(x)$ plot isn't exactly symmetric). ♣

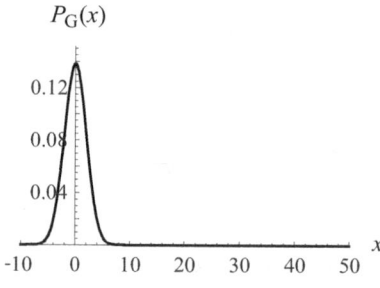

Figure 3.60

57. Gaussian approximation to the Poisson distribution

The first step is to apply Stirling's formula to the $k!$ in the Poisson distribution in Eq. (1.12). This gives

$$P_P(k) \approx \frac{a^k e^{-a}}{k^k e^{-k} \sqrt{2\pi k}}. \qquad (3.324)$$

We claim that the maximum of $P_P(k)$ occurs at a (or technically between $a-1$ and a, but for large a this distinction is inconsequential). We can show this by finding the integer value of k for which both $P_P(k) \geq P_P(k+1)$ and $P_P(k) \geq P_P(k-1)$. $P_P(k)$ is then the maximum, because it is at least as large as the two adjacent

$P_P(k \pm 1)$ values. With $P_P(k) = a^k e^{-a}/k!$, you can quickly show that $P_P(k) \geq P_P(k+1)$ implies $k \geq a - 1$. Similarly, you can show that $P_P(k) \geq P_P(k-1)$ implies $a \geq k$. Combining these two results, we see that the integer value of k that yields the maximum $P_P(k)$ satisfies $a - 1 \leq k \leq a$. The desired value of k is therefore the integer that lies between $a - 1$ and a (or at both of these values if a is an integer).

Since we now know that the maximum of $P_P(k)$ occurs essentially at $k = a$, let's see how $P_P(k)$ behaves near $k = a$. With $x \equiv k - a$ being the number of events relative to a (where a is both exactly the average and approximately the location of the maximum), we have $k = a + x$. In terms of x, Eq. (3.324) becomes

$$P_P(x) \approx \frac{a^{a+x} e^{-a}}{(a+x)^{a+x} e^{-a-x} \sqrt{2\pi(a+x)}}. \tag{3.325}$$

We can cancel the factors of e^{-a}. And we can divide both the numerator and denominator by a^{a+x}. Furthermore, we can ignore the x in the square root, because we'll find below that the x's we're concerned with are small compared with a. The result is

$$P_P(x) \approx \frac{1}{\left(1 + \frac{x}{a}\right)^{a+x} e^{-x} \sqrt{2\pi a}}. \tag{3.326}$$

We'll now apply the approximation from Eq. (1.6) in Problem 53. (As in the solution to Problem 56, the simpler approximation in Eq. (1.5) would yield the wrong answer.) With the a in Eq. (1.6) defined to be x/a here, and with the n defined to be $a + x$, Eq. (1.6) gives

$$\left(1 + \frac{x}{a}\right)^{a+x} \approx \exp\left((a+x)\left(\frac{x}{a}\right) - \frac{1}{2}(a+x)\left(\frac{x}{a}\right)^2\right). \tag{3.327}$$

Multiplying this out and ignoring the small $-x^3/2a^2$ term (because we'll find below that $x \ll a$), we obtain

$$\left(1 + \frac{x}{a}\right)^{a+x} \approx \exp\left(x + \frac{x^2}{2a}\right). \tag{3.328}$$

This is just Eq. (3.315) in the solution to Problem 56, with $n \to a$. Substituting Eq. (3.328) into Eq. (3.326) gives

$$P_P(x) \approx \frac{1}{e^x e^{x^2/2a} e^{-x} \sqrt{2\pi a}}, \tag{3.329}$$

which simplifies to

$$P_P(x) \approx \frac{e^{-x^2/2a}}{\sqrt{2\pi a}} \equiv P_G(x). \tag{3.330}$$

This is the desired Gaussian. If you want to write this result in terms of the actual number k of successes, instead of the number x of successes relative to the average, then using $x = k - a$ gives

$$P_{\rm P}(k) \approx \frac{e^{-(k-a)^2/2a}}{\sqrt{2\pi a}} \equiv P_{\rm G}(k). \tag{3.331}$$

The Poisson distribution (and hence the Gaussian approximation to it) depends on only one parameter, a. And as with the Gaussian approximation to the binomial distribution, the Gaussian approximation to the Poisson distribution is symmetric around $x = 0$ (equivalently, $k = a$).

Fig. 3.61 shows a comparison between the exact $P_{\rm P}(k)$ function in Eq. (1.12) and the approximate $P_{\rm G}(k)$ function in Eq. (3.331). (We've drawn the Gaussians as continuous curves even though only integral values of k are relevant.) The approximation works quite well for $a = 20$ and extremely well for $a = 100$; the curve is barely noticeable behind the dots.

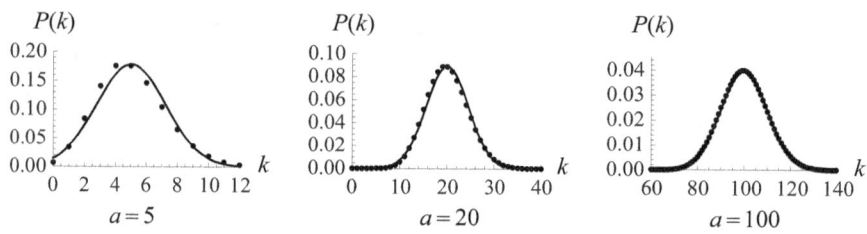

(note different scales on axes)
dots = exact Poisson
solid curve = approximate Gaussian

Figure 3.61

If we compare the Gaussian distribution in Eq. (3.331) with the general Gaussian form from Eq. (3.322) in the solution to Problem 56, we see that the standard deviation is $\sigma = \sqrt{a}$. Since the Poisson distribution depends on only the one parameter a, we already knew that the standard deviation must be a function of a. But it takes some work to show that it equals $\sqrt{a}$.

Note that although $\sqrt{a}$ grows with a, it doesn't grow as fast as a itself. So as a grows, the width of the Poisson-distribution bump (which is proportional to the standard deviation, $\sqrt{a}$) becomes thinner compared with the distance a from the origin to the center of the bump. This is illustrated in Fig. 3.62, where we show the Poisson distributions for $a = 100$ and $a = 1000$. Note the different scales on the axes.

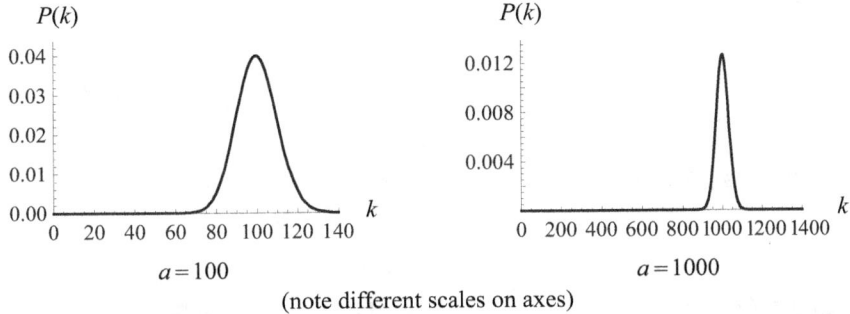

Figure 3.62

Chapter 4

Appendix: Taylor series

Taylor-series approximations are very useful in many problems throughout this book, mainly for checking limiting cases, in particular in situations where a given parameter is small. This appendix gives a brief review of Taylor series.

Note well: Calculus is required if you want to *derive* a Taylor series. However, if you just want to *use* a Taylor series (which is mostly what we do in this book), then algebra is all you need. You just look up the desired series in the list on the following page, and then plug away with whatever algebra is necessary for the task at hand. So although some Taylor-series manipulations might look a bit scary, there's usually nothing more than algebra involved. However, a few problems in this book require you to actually derive a Taylor series, so let's see how that is done.

A Taylor series expresses a given function of x as a series expansion. The general form of a Taylor series, expanded around the point x_0, is

$$f(x) = f(x_0) + f'(x_0)(x - x_0) + \frac{f''(x_0)}{2!}(x - x_0)^2 + \frac{f'''(x_0)}{3!}(x - x_0)^3 + \cdots, \quad (4.1)$$

where the primes denote differentiation. This equality can be verified by taking successive derivatives of both sides of the equation and then setting $x = x_0$. For example, taking the first derivative and then setting $x = x_0$ yields $f'(x_0)$ on the left. And this operation also yields $f'(x_0)$ on the right, because the first term is a constant and gives zero when differentiated, the second term gives $f'(x_0)$, and all the rest of the terms give zero once we set $x = x_0$, because they all contain at least one power of $(x - x_0)$. Likewise, if we take the second derivative of each side and then set $x = x_0$, we obtain $f''(x_0)$ on both sides. And so on for all derivatives. Therefore, since the two functions on each side of Eq. (4.1) are equal at $x = x_0$ and also have their nth derivatives equal at $x = x_0$ for all n, they must in fact be the same function (assuming that they're nicely behaved functions, as we generally assume).

As an example of Eq. (4.1), consider the function $f(x) = \sqrt{1 + x}$. Let's calculate the first few terms of its Taylor series, relative to the origin (that is, with $x_0 = 0$). The function and its first few derivatives are

$$f(x) = (1 + x)^{1/2} \qquad f''(x) = -(1/4)(1 + x)^{-3/2}$$
$$f'(x) = (1/2)(1 + x)^{-1/2} \qquad f'''(x) = (3/8)(1 + x)^{-5/2} \quad (4.2)$$

With $x_0 = 0$, we then have

$$f(0) = 1, \quad f'(0) = 1/2, \quad f''(0) = -1/4, \quad f'''(0) = 3/8. \tag{4.3}$$

Plugging these values into Eq. (4.1) gives the desired Taylor series:

$$\sqrt{1+x} = 1 + (1/2)x - \frac{1/4}{2!}x^2 + \frac{3/8}{3!}x^3 - \cdots$$
$$= 1 + \frac{x}{2} - \frac{x^2}{8} + \frac{x^3}{16} - \cdots. \tag{4.4}$$

As a double check, you can square this result and show that you end up with $1 + x$, up to errors of order x^4. For practice, you can show that the next term in the series is $-(5/128)x^4$. And then you can verify that this correctly gets rid of the x^4 term in the squaring operation, pushing the error down to order-x^5. Indeed, demanding that the square equals $1 + x$, up to errors of order x^k, for successively increasing values of k, is a perfectly valid way of deriving the Taylor series, step by step. No derivatives required. (This works for a simple square-root function, but many functions are more complicated and require you to calculate the derivatives in Eq. (4.1).)

Some specific Taylor series that often come up are listed below. They are all expanded around $x = 0$. That is, $x_0 = 0$ in Eq. (4.1). They are all derivable via Eq. (4.1), but sometimes there are quicker ways of obtaining them. For example, Eq. (4.6) is most easily obtained by taking the derivative of Eq. (4.5), which itself is just the sum of a geometric series.

$$\frac{1}{1+x} = 1 - x + x^2 - x^3 + \cdots \tag{4.5}$$

$$\frac{1}{(1+x)^2} = 1 - 2x + 3x^2 - 4x^3 + \cdots \tag{4.6}$$

$$\ln(1+x) = x - \frac{x^2}{2} + \frac{x^3}{3} - \cdots \tag{4.7}$$

$$e^x = 1 + x + \frac{x^2}{2!} + \frac{x^3}{3!} + \cdots \tag{4.8}$$

$$\cos x = 1 - \frac{x^2}{2!} + \frac{x^4}{4!} - \cdots \tag{4.9}$$

$$\sin x = x - \frac{x^3}{3!} + \frac{x^5}{5!} - \cdots \tag{4.10}$$

$$\sqrt{1+x} = 1 + \frac{x}{2} - \frac{x^2}{8} + \cdots \tag{4.11}$$

$$\frac{1}{\sqrt{1+x}} = 1 - \frac{x}{2} + \frac{3x^2}{8} + \cdots \tag{4.12}$$

$$(1+x)^n = 1 + nx + \binom{n}{2}x^2 + \binom{n}{3}x^3 + \cdots \tag{4.13}$$

(For the trig functions here, x is measured in radians, not degrees.) Each of these series has a range of validity, that is, a "radius of convergence." For example, the series for e^x is valid for all x, while the series for $1/(1+x)$ is valid for $|x| < 1$. The various ranges won't particularly concern us, because whenever we use one of the above Taylor series, we will assume that x is small (much smaller than 1). In this case, all of the series are valid.

The above list might seem a little intimidating, but in most situations there is no need to include terms beyond the first-order term in x. For example, $\sqrt{1+x} \approx 1+x/2$ is usually a good enough approximation. (The square of it equals $1 + x$, up to errors of order x^2.) The smaller x is, the better this approximation is, because any term in the expansion is smaller than the preceding term by a factor of order x. We'll have more to say about how many terms to include, at the end of the appendix.

As mentioned above, you often don't need to worry about taking derivatives and rigorously deriving a Taylor series. You just take it as given, which means that if you haven't studied calculus yet, that's no excuse for not using a Taylor series! All you need to do is refer to the above list for the series you're interested in. If you want to check that a particular series is believable, you can just use your calculator. For example, consider what e^x looks like if x is a very small number, say, $x = 0.001$. Your calculator (or a computer, if your want more digits) will tell you that

$$e^{0.001} = 1.001\,000\,500\,166\,7\ldots \qquad (4.14)$$

This can be written more informatively as

$$\begin{aligned} e^{0.001} &= 1.0 \\ &+ 0.001 \\ &+ 0.000\,000\,5 \\ &+ 0.000\,000\,000\,166\,7\ldots \\ &= 1 + (0.001) + \frac{(0.001)^2}{2!} + \frac{(0.001)^3}{3!} + \cdots. \end{aligned} \qquad (4.15)$$

This last line agrees with the form of the Taylor series for e^x in Eq. (4.8). If you made x smaller (say, 0.00001), then the same pattern would appear, but just with more zeros between the numbers than in Eq. (4.14). If you kept more digits in Eq. (4.14), you could verify the $x^4/4!$ and $x^5/5!$, etc., terms in the e^x Taylor series. But things aren't quite as obvious for these terms, because we don't have all the nice zeros as we do in seven of the first nine digits in Eq. (4.14).

Note that the lefthand sides of the Taylor series in the above list involve 1's and x's. So how do we make an approximation to an expression of the form, say, $\sqrt{N+x}$, where x is small? We could of course use the general Taylor-series expression in Eq. (4.1) to generate the series from scratch by taking derivatives. But we can save ourselves some time by making use of the similar-looking $\sqrt{1+x}$ series in Eq. (4.11). We can turn the N into a 1 by factoring out an N from the square root, which gives $\sqrt{N}\sqrt{1+x/N}$. Having generated a 1, we can now apply Eq. (4.11), with the only modification being that the small quantity x that appears in that equation is replaced by the small quantity x/N. This gives (to first order in x)

$$\sqrt{N+x} = \sqrt{N}\sqrt{1+\frac{x}{N}} \approx \sqrt{N}\left(1 + \frac{1}{2}\frac{x}{N}\right) = \sqrt{N} + \frac{x}{2\sqrt{N}}. \qquad (4.16)$$

You can quickly verify that this expression is valid to first order in x by squaring both sides. As a numerical example, if $N = 100$ and $x = 1$, then this approximation gives $\sqrt{101} \approx 10 + 1/20 = 10.05$, which is very close to the actual value of $\sqrt{101} \approx 10.0499$.

EXAMPLE 1 (CALCULATING A SQUARE ROOT): Use the Taylor series $\sqrt{1+x} \approx 1 + x/2 - x^2/8$ to produce an approximate value of $\sqrt{5}$. How much does your answer differ from the actual value?

SOLUTION: We'll first write 5 as $4 + 1$, because we know what the square root of 4 is. However, we can't apply the given Taylor series with $x = 4$, because we need x to be small. We must first factor out a 4 from the square root, so that we have an expression of the form $\sqrt{1+x}$, where x is small. Using $\sqrt{1+x} \approx 1 + x/2 - x^2/8$ with $x = 1/4$ (not 4!), we obtain

$$\sqrt{5} = \sqrt{4+1} = 2\sqrt{1+1/4} \approx 2\left(1 + \frac{1/4}{2} - \frac{(1/4)^2}{8}\right)$$

$$= 2\left(1 + \frac{1}{8} - \frac{1}{128}\right) \approx 2.2344. \tag{4.17}$$

The actual value of $\sqrt{5}$ is about 2.2361. The approximate result is only 0.0017 less than this, so the approximation is quite good (the percentage difference is only 0.08%). Equivalently, the square of the approximate value is 4.9924, which is very close to 5. If you include the next term in the series, which is $+x^3/16$ from Eq. (4.4), the result is $\sqrt{5} \approx 2.2363$, with an error of only 0.01%. By keeping a sufficient number of terms, you can produce any desired accuracy.

When trying to determine the square root of a number that isn't a perfect square, you could of course just guess and check, improving your guess on each iteration. But a Taylor series (calculated relative to the closest perfect square) provides a systematic method that doesn't involve guessing. ∎

EXAMPLE 2 (LIMIT OF A QUOTIENT): What does $(e^x - 1)/x$ equal, in the $x \to 0$ limit?

SOLUTION: As $x \to 0$, both the numerator and denominator of $(e^x - 1)/x$ go to zero, so we obtain $0/0$, which is undefined (and could be anything). But if we use the Taylor series $e^x \approx 1 + x$, we can write

$$\frac{e^x - 1}{x} \approx \frac{(1+x) - 1}{x} = \frac{x}{x} = 1. \tag{4.18}$$

And since the $x \to 0$ limit of the number 1 is just 1, of course, the desired limit is 1. You can check this with your calculator.

If you're worried that additional terms in the Taylor series for e^x might mess things up, you can include them and write

$$\frac{e^x - 1}{x} = \frac{(1 + x + x^2/2 + x^3/6 + \cdots) - 1}{x} = 1 + x/2 + x^2/6 + \cdots. \tag{4.19}$$

In the $x \to 0$ limit, all the terms involving x go to zero, so we're left with only the 1.

A common procedure for dealing with 0/0 expressions is l'Hôpital's rule, which you may be familiar with. The rule involves taking derivatives (you can look up the details). If you instead use the above Taylor-series method, you're really doing the same thing in the end; the recipe in Eq. (4.1) for creating a Taylor series involves taking derivatives. The Taylor method is effectively a proof of the l'Hôpital method, as you can show. ∎

When making a Taylor-series approximation, how do you know how many terms in the series to keep? For example, if the exact answer to a given problem takes the form of $e^x - 1$, then the Taylor series $e^x \approx 1 + x$ tells us that our answer is approximately equal to x. You can check this by picking a small value for x (say, 0.01) and plugging it into your calculator. This approximate form makes the dependence on x (for small x) much more transparent than the original expression $e^x - 1$ does.

But what if our exact answer had instead been $e^x - 1 - x$? The Taylor series $e^x \approx 1 + x$ would then yield an approximate answer of zero. And indeed, the answer *is* approximately zero. However, when making approximations, it is generally understood that we are looking for the *leading-order* term in the answer (that is, the smallest power of x with a nonzero coefficient). If our approximate answer comes out to be zero, then that means we need to go (at least) one term further in the Taylor series, which means $e^x \approx 1 + x + x^2/2$ in the present case. Our approximate answer is then $x^2/2$. (You should check this by letting $x = 0.01$.) Similarly, if the exact answer had instead been $e^x - 1 - x - x^2/2$, then we would need to go out to the $x^3/6$ term in the Taylor series for e^x.

You should be careful to be consistent with the powers of x you use. If the exact answer is, say, $e^x - 1 - x - x^2/3$, and if you use the Taylor series $e^x \approx 1 + x$, then you will obtain an approximate answer of $-x^2/3$. This is incorrect, because it is inconsistent to pay attention to the $-x^2/3$ term in the exact answer while ignoring the corresponding $x^2/2$ term in the Taylor series for e^x. Including both terms gives the correct approximate answer of $x^2/6$.

So what is the answer to the above question: How do you know how many terms in the series to keep? Well, the answer is that before you do a (perhaps messy) calculation, there's really no way of knowing how many terms to keep. The optimal strategy is probably to just hope for the best and start by keeping only the term of order x. This will often be sufficient. But if you end up with a result of zero, then you can go to order x^2, and so on. Of course, you could play it safe and always keep terms up to, say, fourth order. But that is invariably a poor strategy, because you will probably never actually need to go out that far in a series, meaning that the horrendous algebra you just inflicted upon yourself was all for naught.

www.ingramcontent.com/pod-product-compliance
Lightning Source LLC
Chambersburg PA
CBHW062350220526
45472CB00008B/1765